美味西餐

新东方烹饪教育 组编

中国人民大学出版社
·北京·

编委会名单

编委会主任　金晓峰
编委会副主任　汪　俊
编　　　委　谭荣华　温杰峰　冯吉香　王海森
拍　　　摄　吴礼洋

C O N T E N T S

目 录

CHAPTER 1
沙拉

2	凯撒沙拉
4	希腊沙拉
6	番茄奶酪沙拉
8	华尔道夫沙拉
10	厨师沙拉

CHAPTER 2
头盘

14	烟熏三文鱼配牛油果泥
16	鹅肝慕斯
18	煎金枪鱼配坚果碎
20	海鲜鸡尾杯
22	煎扇贝配南瓜泥
24	培根肠仔卷
26	芝士焗土豆
28	法式凤尾虾

CHAPTER 3
蛋类

32	水波蛋
34	带壳溏心蛋
36	蛋黄酱水煮蛋
38	法式番茄炖蛋

CHAPTER 4
小食

42　俱乐部三明治
44　金枪鱼帕尼尼
46　牛肉汉堡
48　热狗包配薯条佐热狗酱
50　炸鱼柳配太太沙司
52　可丽饼
54　橄榄面包
56　橙味法式芝多士
58　香酥鸡米花

CHAPTER 5
汤类

62　西班牙冷汤
64　意大利蔬菜汤
66　奶油蘑菇汤
68　马赛鱼汤
70　罗宋汤
72　鲜虾浓汤

CHAPTER 6
配菜

76　芝士舒芙蕾
78　土豆泥
80　蒜香意面
82　洋葱培根薯角
84　香草烤杂蔬

CONTENTS 目录

CHAPTER 7
装饰配菜

- 88　墨鱼脆片
- 90　芝士薄片
- 92　菠菜芝士条
- 94　墨鱼西米薄脆

CHAPTER 8
主菜

- 98　惠灵顿牛排
- 100　香煎三文鱼
- 102　意式焗龙虾
- 104　油封香草鸭
- 106　芝士鸡肉卷
- 108　主妇鳕鱼
- 110　黄油鸡卷
- 112　宝云苏羊排
- 114　法式焗蜗牛
- 116　海鲜意面
- 118　香草面包糠焗青口
- 120　蟹肉土豆饼

CHAPTER 9
东南亚菜

- 124　越南春卷
- 126　泰式牛肉色拉
- 128　冬阴功汤
- 130　拉萨汤
- 132　印尼炒饭
- 134　马来姜蓉虾
- 136　桑巴烤鱼

第1章
沙拉

凯撒沙拉
Caesar Salad

菜肴特点
(Dish Characteristic)

鲜咸可口，营养丰富
Fresh and salty, rich in nutrition

　　凯撒沙拉被称为"沙拉之王"。关于这道菜流传的说法是：有一天餐厅快耗尽厨房的原料，这时主厨凯撒·卡狄尼用仅剩的原料制作出凯撒沙拉。
　　Caesar salad is known as the king of salad. One day, the restaurant almost ran out of ingredients, Chef Caesar Cardini made Caesar salad with what he had.

原料 (Ingredients)

罗马生菜	Romaine Lettuce	200g
蛋黄酱	Mayonnaise	500ml
面包丁	Croutons	10g
盐	Salt	适量
培根	Bacon	20g
胡椒	Pepper	适量
大蒜	Garlic	5g
醍鱼柳	Anchovy	2条
帕玛森芝士粉	Parmesan Cheese Powder	10g
鸡胸肉	Chicken breast	50g

温馨提示 (Kindly Reminder)

鸡胸肉可选择放，也可选择不放。
(Chicken breast meat is optional.)

制作过程 (Method)

① 将所有原料准备齐全。
(Prepare all the ingredients.)

② 将罗马生菜洗净，切成 2cm 左右的长段。在蛋黄酱中加入大蒜末、醍鱼柳碎、帕玛森芝士粉，搅拌均匀成酱汁。
(Wash Romaine Lettuce and cut them into 2cm length. For preparing the salad dressing, mix mayonnaise with grind garlic, chopped anchovy and Parmesan Cheese Powder.)

③ 将培根放入煎锅中煎熟。
(Pan-fry the bacon.)

④ 将鸡胸肉用基本味腌制（盐、胡椒）后煎至上色，放入烤箱（180℃，烤8分钟），取出冷却后切成片。
(Marinade chicken breast, put it into oven at 180℃ for 8 minutes, take out from oven and cool it, cut them into slices.)

⑤ 将做好的酱汁放入色拉中搅拌均匀。
(Mix the dressings with salad and stir well.)

⑥ 最后撒上培根片、面包丁、芝士粉，装盘即可。
(Finally sprinkle with bacon slices, diced bread and cheese powder, and it is ready to serve.)

①

②

③

④

⑤

⑥

希腊沙拉
Greek Salad

菜肴特点
(Dish Characteristic)

营养丰富，色泽鲜艳
A fresh and colorful salad with lots of healthy ingredients

希腊沙拉

原料 (Ingredients)

菲达芝士	Feta Cheese	10g
番茄	Tomato	5g
洋葱	Onion	10g
柠檬	Lemon	50g
黄瓜	Cucumber	10g
欧芹	Parsley	10g
红酒醋	Red wine vinegar	10g
生菜	Lettuce	10g
大蒜	Garlic	5g
彩椒	Pepper	3g
黑橄榄	Black olive	5g
橄榄油	Olive oil	5g
阿里根奴	Oregano	5g
盐	Salt	3g
胡椒	Pepper	5g

温馨提示 (Kindly Reminder)

酱汁不可提前放，否则生菜容易蔫掉。
(Put the salad dressing at the end, otherwise the lettuces will get watery.)

制作过程 (Method)

① 将所有原料准备齐全。
(Prepare all the ingredients.)

② 将黄瓜、彩椒、洋葱、番茄切丁，黑橄榄切小圈。
(Dice the cucumber, peppers, onion, tomato, cut black olive into small rings.)

③ 将菲达芝士切丁，将原料中的蔬菜(番茄、洋葱、黄瓜、生菜、彩椒)加工成型后备用。
(Dice the Feta Cheese into cubes, get all washed up vegetables ready, set aside.)

④ 将大蒜、欧芹切末，加入柠檬汁、红酒醋、阿里根奴、橄榄油、盐和胡椒，做成红酒醋汁。
(Chop garlic and parsley, add lemon juice, red wine vinegar, Oregano, olive oil, salt and pepper to make red wine vinegar dressing.)

⑤ 将红酒醋汁淋在蔬菜色拉里拌匀。
(Pour red wine vinegar dressing over salad, stir well.)

⑥ 最后，加入欧芹、黑橄榄装盘即可。
(Add parsley and black olive, it is ready to serve.)

❶

❷

❸

❹

❺

❻

番茄奶酪沙拉
Tomato with Mozzarella Cheese Salad

菜肴特点
(Dish Characteristic)

奶香浓郁，柔软弹滑，色泽鲜艳
Rich milk flavor, soft and smooth, bright colour

番茄奶酪沙拉

原料 (Ingredients)

番茄	Tomato	100g
马苏里拉芝士	Mozzarella Cheese	100g
罗勒叶	Basil Leaves	5g
松子仁	Pine nuts	150ml
大蒜	Garlic	10g
帕玛森芝士粉	Parmesan Cheese powder	10g
橄榄油	Olive oil	5g
盐	Salt	4g
胡椒	Pepper	5g

温馨提示 (Kindly Reminder)

（1）番茄和芝士切片厚薄一致。
(Tomato and cheese slice should be cut into the same size and thickness.)

（2）罗勒酱最好现做现打，可以保持鲜艳的绿色。
(Freshly made Basil Sauce can keep bright green.)

制作过程 (Method)：

① 将所有原料准备齐全。
(Prepare all the ingredients.)

② 将番茄切成1厘米厚的薄片。
(Cut the tomatoes into slice of 1cm thick.)

③ 将马苏里拉芝士切成1厘米厚的薄片。
(Cut the Mozzarella Cheese into slice of 1cm thick.)

④ 将大蒜、罗勒叶、帕玛森芝士粉、松子仁、盐、胡椒和橄榄油混合，放入搅拌机打成罗勒酱。
(Add garlic, Basil Leaves, Parmesan Cheese Powder, pine nuts, salt, pepper and olive oil into the blender, mixing well to make the Basil Sauce.)

⑤ 将罗勒酱、番茄片、芝士片摆盘。
(Put Basil Sauce, tomato slices, cheese slices on the plate.)

⑥ 撒上松子仁，配上罗勒叶装盘即可。
(Sprinkle with pine nuts and serve with Basil Leaves.)

❶

❷

❸

❹

❺

❻

华尔道夫沙拉
Waldorf Salad

菜肴特点
(Dish Characteristic)

酸甜可口，营养丰富
Sour and sweet, rich in nutrition

　　华尔道夫沙拉具有100多年的历史，最初原料中只有西芹、苹果及蛋黄酱，后来又加入了核桃仁，形成苹果、西芹、核桃仁的黄金搭配，带来口感上的升华，妙不可言！
　　Waldorf Salad has a history of more than 100 years. The original recipe was just celery, apple and mayonnaise. Then chopped walnut had been added, and it tastes well with apple and celery. Too wonderful to express in words!

原料 (Ingredients)

苹果	Apple	50g
葡萄干	Raisin	10g
西芹	Celery	20g
盐	Salt	5g
核桃仁	Walnut	5g
胡椒	Pepper	5g
蛋黄酱	Mayonnaise	10g
柠檬汁	Lemon juice	4g

温馨提示 (Kindly Reminder)

（1）西芹和苹果丁需切得大小一致。
(Celery and apple diced need to be cut into the same size.)

（2）西芹需要刨去外面老筋。
(The celery needs to be peeled.)

制作过程 (Method)

① 将所有原料准备齐全。
(Prepare all the ingredients.)

② 将西芹去筋，洗净；苹果去皮，都切成 1cm 大小的丁。
(Trim the celery ; peel the apples and cut them into 1cm chunks.)

③ 将核桃仁放入烤箱，用 180℃烤 8 分钟，掰成小块。
(Place walnuts on a baking tray, bake at 180℃ for 8 minutes, roughly chop into small pieces.)

④ 将切好的苹果丁和西芹丁混合，加入蛋黄酱、盐、胡椒、柠檬汁拌匀。
(Place the diced apple and celery, drizzle over the mayonnaise,salt,pepper,lemon juice and toss well.)

⑤ 撒上葡萄干和核桃。
(Sprinkle with raisins and walnuts.)

⑥ 放入装饰物装盘即可。
(Add garnishes, then serve.)

❶

❷

❸

❹

❺

❻

厨师沙拉
Chef Salad

菜肴特点
(Dish Characteristic)

清爽可口，营养丰富
A fresh and delicious salad with lots of healthy ingredients

厨师沙拉

原料 (Ingredients)

红圆椒	Red pepper	20g
小番茄	Cherry tomato	适量
黄圆椒	Yellow pepper	20g
西芹	Celery	10g
青圆椒	Green pepper	20g
熟鸡蛋	Boiled egg	适量
洋葱	Onion	10g
车打芝士片	Cheddar Cheese	20g
生菜	Lettuce	25g
玉米	Corn	10g
火腿	Ham	20g
小萝卜	Radish	5g

 温馨提示 (Kindly Reminder)

可根据需要配上油醋汁或色拉酱。
(You may put the vinegar dressing or mayonnaise if necessary.)

制作过程 (Method)

① 将所有原料准备齐全并洗净。
(Prepare all the ingredients and get all washed.)

② 将小番茄洗净后切成三角形。
(Cut cherry tomatoes into triangle shape.)

③ 将彩椒切成三角形。
(Cut pepper into triangle shape.)

④ 将熟鸡蛋、火腿、芝士切成小片,玉米煮熟后备用。
(Cut boiled eggs, ham and cheese into small pieces, corn bailed and set aside.)

⑤ 将生菜、洋葱、小萝卜、西芹切片。
(Slice the lettuce, onion, radish and celery.)

⑥ 最后加入装饰物装盘即可。
(Place the garnishes and it is ready to serve.)

①

②

③

④

⑤

⑥

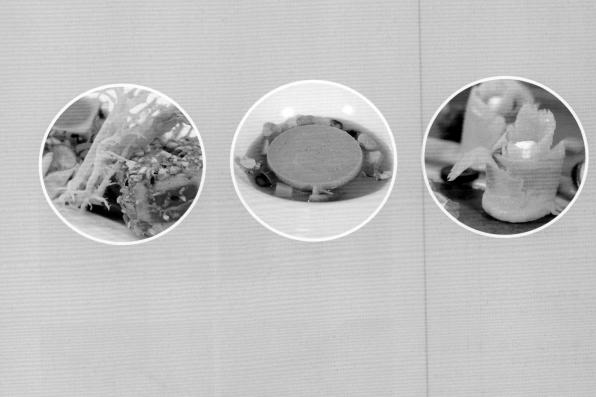

烟熏三文鱼配牛油果泥

Smoked Salmon with Avocado Puree

菜肴特点
（Dish Characteristic）

口感咸鲜，酸甜可口，具有烟熏味，色泽鲜艳
The palate is salty and fresh, sweet and sour, with a smoky flavor and bright colour

烟熏三文鱼配牛油果泥

原料 (Ingredients)

烟熏三文鱼	Smoked salmon	200g
牛油果	Avocado	50g
柠檬	Lemon	10g
糖	Sugar	10g
淡奶油	Cream	10g
黑橄榄	Black olive	5g
水瓜柳	Capers	5g
小萝卜	Radish	5g

温馨提示 (Kindly Reminder)

酸奶油可用淡奶油加柠檬汁制作而成。
(Sour cream can be made of light cream and lemon juice.)

制作过程 (Method)

① 将所有原料准备齐全。
(Prepare all the ingredients.)

② 将牛油果去核去皮，切成小块。
(Peel the avocado skin and seeds, dice the avocado into small cubes.)

③ 在牛油果中加入糖和柠檬汁、水瓜柳，放入打汁机打成泥。
(Put avocado into the blender, add sugar lemon juice and capers.)

④ 将烟熏三文鱼卷成花型，倒入牛油果泥。
(Make smoked salmon roll into shape, then pour in puree.)

⑤ 将所有原料（如图）都加工成型备用。
(Prepare all the ingredients as the picture shows.)

⑥ 最后挤上酸奶油，加入黑橄榄，装盘即可。
(Place the sour cream and black olive, then serve with the dish.)

❶

❷

❸

❹

❺

❻

鹅肝慕斯
Foie Gras Mousse

菜肴特点
(Dish Characteristic)

口感丝滑
This foie gras mousse is silky and creamy

原料 (Ingredients)

鹅肝	Foie gras	200g
芒果	Mango	30g
淡奶油	Cream	20g
草莓	Strawberry	10g
蓝莓	Blueberry	5g
糖	Sugar	10g
红酒	Red wine	10g
盐	Salt	5g
胡椒	Pepper	4g

温馨提示 (Kindly Reminder)

鹅肝必须选用新鲜的。
(Fresh foie gras is necessary.)

制作过程 (Method)

① 将所有原料准备齐全。
（Prepare all ingredients.）

② 将鹅肝去筋，然后用盐、胡椒及红酒腌制。
（Trim the tender part of foie gras and marinate with salt, pepper and red wine.）

③ 将腌好的鹅肝放入真空袋，然后放入温水中煮30分钟。
（Vacuum seal the foie gras in a cooking bag, place the foie gras in warm water for 30 minutes.）

④ 在煮熟冷却后的鹅肝中加入淡奶油，然后放入模具中，再放进冰箱冷藏一小时以上。
（Add whipping cream to the foie gras, pour it in the mold and refrigerate for an hour.）

⑤ 将芒果和糖制成芒果酱垫底，放入鹅肝慕斯。
（Place the mango jam at the bottom of foie gras mousse.）

⑥ 最后放上草莓、蓝莓，装盘即可。
（Decorate with strawberry and blueberry, then serve.）

❶

❷

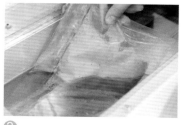
❸

❹

❺

❻

煎金枪鱼配坚果碎
Tuna with Mustard Nutty Crust

菜肴特点
(Dish Characteristic)

色泽鲜艳，外脆里嫩
Such a colorful dish, it tastes crispy outside and soft inside

金枪鱼富含丰富的维生素、矿物质和少许微量元素，是女性美容养颜、减肥的健康食品，有防止动脉硬化、保护肝脏的作用。

Tuna is rich in vitamins, minerals and trace elements, It is a healthy food for women for beautifying, nourishing and losing weight. It can prevent arteriosclerosis and protect the liver.

煎金枪鱼配坚果碎

原 料 (Ingredients)

金枪鱼	Tuna	150g
黑芝麻	Black Sesame	10g
开心果碎	Pistachio	10g
芥末	Mustard	50g
杏仁碎	Nutmeg	10g
浓缩黑醋	Balsamic	20g
草莓	Strawberry	10g
盐	Salt	4g
胡椒	Pepper	5g

温馨提示 (Kindly Reminder)

（1）金枪鱼必须选用新鲜的。
(Fresh tuna is necessary.)

（2）坚果碎和黑芝麻也可以用烤箱烤熟。
(Place the crushed nuts and Black Sesame seeds on the baking tray, and bake it until golden.)

制作过程 (Method)

① 将所有原料准备齐全。
(Prepare all the ingredients.)

② 将金枪鱼切成型后用盐、胡椒腌制入味，然后放入热煎锅中煎上色。
(Cut the tuna in strip, marinate with salt and pepper. Heat the pan, then pan-fry until it changes dark red colour.)

③ 将开心果碎、杏仁碎、黑芝麻炒熟备用。
(Stir-fry the pistachio, nutmeg and the Black Sesame seeds.)

④ 将金枪鱼外面涂上芥末，然后裹上混合的坚果碎和黑芝麻。
(Coat the tuna with mustard and the mixture of crushed nuts and Black Sesame seeds.)

⑤ 将金枪鱼改刀成型。
(Using a sharp knife, slice the tuna into squares.)

⑥ 最后用浓缩黑醋画出轮廓，加入草莓等装饰物，装盘即可。
(Outline with balsamic, and place the garnish, then ready to serve.)

①

②

③

④

⑤

⑥

海鲜鸡尾杯
Seafood Cocktail

菜肴特点
（Dish Characteristic）

酸甜可口，微辣，色泽鲜艳
Sweet and sour, slightly spicy, bright in colour

海鲜鸡尾杯

原料 (Ingredients)

虾	Shrimp	100g
鱿鱼	Squid	100g
番茄酱	Tomato sauce	20g
白兰地	Brandy	5g
美国辣酱汁	Tabasco	3g
李派林喼汁	Worcestershire	5g
柠檬	Lemon	5g
盐	Salt	3g
胡椒	Pepper	2g
生菜	Lettuce	4g

温馨提示 (Kindly Reminder)

（1）海鲜必须选用新鲜的。
(Seafood must be fresh.)

（2）装盘时选用鸡尾酒杯比较合适。
(Cocktail cups are appropriate for serving.)

制作过程 (Method)

① 将所有原料准备齐全。
(Prepare all the ingredients.)

② 将鱿鱼进行刀工处理。
(Cut all the squid into the shape you want.)

③ 将虾放入沸水锅中焯水至成熟。
(Put the shrimp in a boiling pot and blanch until it is cooked.)

④ 调制鸡尾汁：将番茄酱、李派林喼汁、美国辣酱汁、柠檬汁、白兰地、盐、胡椒混合后搅拌均匀。
(Cocktail sauce: Mix tomato sauce, Worcestershire, lemon juice, Tabasco, Brandy, salt, pepper well.)

⑤ 在海鲜中拌入鸡尾汁调制入味。
(Add cocktail sauce to cooked seafood to taste.)

⑥ 最后装盘时配上生菜和柠檬皮末。
(Place the lettuce and lemon zest over the seafood then it's ready to serve.)

❶

❷

❸

❹

❺

❻

煎扇贝配南瓜泥
Pan Fried Scallop with Pumpkin Puree

菜肴特点
（Dish Characteristic）

口感咸鲜，奶香浓郁，色泽艳丽
The palate is salty, fresh, creamy and bright in colour

# 煎扇贝配南瓜泥

原料 (Ingredients)

扇贝	Scallop	200g
节瓜	Zucchini	10g
南瓜	Pumpkin	100g
黄油	Butter	10g
胡萝卜	Carrot	5g
玉米	Corn	5g
胡椒	Pepper	5g
盐	Salt	6g

温馨提示 (Kindly Reminder)

（1）扇贝可选用冰鲜的大扇贝。
(Large frozen scallop can be used.)

（2）南瓜可选择烤熟，也可选择蒸熟。
(The pumpkin can be roasted or steamed.)

制作过程 (Method)

① 将所有原料准备齐全。
(Prepare all the ingredients.)

② 将南瓜改刀后放入烤箱，用180℃烤1小时左右，然后打成泥。
(Put pumpkin into the oven and roast at 180℃ for 1 hour, then puree.)

③ 将节瓜、胡萝卜切成长条片，焯水待用。
(Put the zucchini, carrot into strips and blanch them.)

④ 起一个热锅，加入黄油、盐和胡椒，将扇贝腌制后煎熟，加入节瓜、胡萝卜、玉米煎入味。
(Heat up a pan, add butter, salt and pepper. Marinate scallops and pan-fry it, then add zucchini, carrot and corn.)

⑤ 最后装盘即可。
(Place the garnish and ready to serve.)

❶

❷

❸

❹

❺

培根肠仔卷
Bacon with Sausage Roll

菜肴特点
（Dish Characteristic）

咸鲜可口，造型美观，花纹清晰
Salty and delicious, beautiful modeling, decorative pattern is clear

原料 (Ingredients)

培根	Bacon	50g
烧烤酱	Barbecue sauce	20g
香肠	Sausage	100g
番茄酱	Tomato sauce	30g
蜂蜜	Honey	5g
李派林喼汁	Worcestershire	5g
大蒜	Garlic	10g
生菜	Lettuce	10g
色拉油	Oil	10g
盐	Salt	3g
胡椒	Pepper	2g

温馨提示 (Kindly Reminder)

还可以搭配其他酱汁，如泰式甜辣酱等。
(The dish can serve with Thai sweet and spicy sauce.)

制作过程 (Method)

①将所有原料准备齐全。
(Prepare all the ingredients.)

②将香肠切段，然后在香肠中间裹上培根并开十字花刀。
(Cut the sausage into smaller sausage. Wrap one half rasher of bacon around each sausage, make a cross on each sausage.)

③用竹签将培根和香肠穿在一起。
(Use a bamboo stick to tie up bacon and sausage together.)

④将培根香肠卷放入五成热的色拉油中炸至金黄色。
(Heat oil in a pan. Make sure the oil is heated enough and put the sausage roll in it till golden brown.)

⑤将大蒜炒香后，加入烧烤酱、番茄酱、李派林喼汁、蜂蜜、盐、胡椒调味成酱汁备用。
(Stir-fry the garlic until fragrant, add barbecue sauce, tomato sauce, Worcestershire, honey, season with salt and pepper, set aside.)

⑥最后装盘时配上生菜及其他装饰物即可。
(Garnish with lettuce and other garnishes, then it is ready to serve.)

❶

❷

❸

❹

❺

❻

芝士焗土豆
Potato Gratin with Cheese

菜肴特点
（Dish Characteristic）

奶香浓郁，柔滑软糯，色泽金黄
A thick milky flavor, soft and smooth, golden colour

芝士焗土豆

原料 (Ingredients)

马铃薯	Potato	1个
胡萝卜	Carrot	适量
香草束	Herbs	1个
淡奶油	Cream	150ml
豆蔻	Nutmeg	适量
节瓜	Zucchini	适量
牛奶	Milk	500ml
葛利亚芝士粉	Gruyere Cheese	100g
盐	Salt	适量
胡椒	Pepper	适量

温馨提示 (Kindly Reminder)

（1）马铃薯切片厚薄一致。
(Potato slice should be cut into the same size.)

（2）芝士要撒均匀，菜肴表面需焗成金黄色。
(The cheese is spread evenly, bake it into golden brown.)

制作过程 (Method)

① 将所有原料准备齐全。
(Prepare all the ingredients.)

② 将马铃薯去皮，切成3厘米薄片。将牛奶倒入锅内，烧开，加入香草束、豆蔻并用盐和胡椒调味；放入马铃薯片，烧开，然后用小火煮8分钟，捞出马铃薯沥干。
(Peel the potato skin, slice the potato into 3cm thickness. Bring slice potato and milk to a boil, add herbs, nutmeg and seasoning with salt and pepper; turn into low heat and simmer for 8 minutes, then drain it.)

③ 将马铃薯叠放在烤盘内，每一层都放入盐、胡椒、豆蔻，最后，在表面撒上磨碎的葛利亚芝士粉，用180℃烤8分钟,烤至表面呈金黄色。
(Stack the potatoes into layer, put salt, pepper and nutmeg on each layer, and then sprinkle with Gruyere Cheese. Put in the oven and bake for 8 minutes at 180 degrees at the cheese turn into golden yellow.)

④ 将节瓜和胡萝卜加工成型后进行焯水处理备用。
(Blanching zucchini and carrot after they are processed.)

❶

❷

❸

❹

❺

❻

⑤ 在牛奶中加入淡奶油，烧开至浓稠，最后用芝士调味。
(Cook the cream sauce with low heat until it boils and thickens, then add some cheese into the sauce.)

⑥ 用芝士垫底，放上马铃薯、配菜装盘即可。
(Place the cream cheese in the bottom, put the potatoes over the cream cheese, and ready to serve.)

法式凤尾虾

French Phoenix-tailed Prawns

菜肴特点
(Dish Characteristic)

色泽鲜艳，外脆里嫩，鲜甜可口
Bright in colour, crisp outside and soft inside, fresh and sweet

原料 (Ingredients)

大虾	Prawn	200g
鸡蛋	Egg	50g
柠檬	Lemon	5g
面粉	Flour	150g
白兰地	Brandy	5g
面包糠	Breadcrumb	150g
蛋黄酱	Mayonnaise	100g
盐	Salt	2g
胡椒	Pepper	3g

温馨提示 (Kindly Reminder)

大虾选用新鲜的口感较好。
(Fresh prawns are preferred.)

制作过程 (Method)

① 将所有原料准备齐全。
(Prepare all the ingredients.)

② 将大虾去肠，开背。
(Use a paring knife to make a slit along the shrimp's back. Pull the vein out gently with your fingers.)

③ 将大虾用白兰地、盐、胡椒腌制入味。
(Marinate the prawns with Brandy, salt and pepper.)

④ 在大虾的背部挤上蛋黄酱。
(Squeeze some mayonnaise at the back of the shrimp.)

⑤ 将虾肉"过三关"（面粉、鸡蛋、面包糠）后用五成油温炸至成熟。
(The first layer of shrimp is flour wrapped, second layer of egg wrapped, and third layer of breadcrumb, fry at 180 degrees or so.)

⑥ 最后配上柠檬等装饰物装盘即可。
(Serve fried shrimp with lemons and other garnishes.)

❶

❷

❸

❹

❺

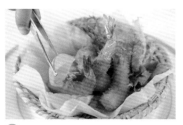
❻

第3章 蛋类

水波蛋
Poach Egg

菜肴特点
(Dish Characteristic)

营养丰富，口感软嫩
Rich in nutrition, soft and tender taste

在水波蛋的做法中，普遍会在沸水中加入白醋，这样能够加快蛋清的凝固速度，而且煮出的鸡蛋清会更加白皙、嫩滑。

From a lot of poached eggs recipe, usually it adds white vinegar to boiling water, it can accelerate the speed of egg white solidification and the egg will be more white and tender.

水波蛋

原料 (Ingredients)

鸡蛋	Egg	50g
白醋	White vinegar	10g

温馨提示 (Kindly Reminder)

（1）鸡蛋必须选用新鲜的。
(Egg must be fresh.)

（2）水波蛋可以搭配荷兰汁佐以食用。
(Poach egg can be served with hollanders.)

制作过程 (Method)

① 将所有原料准备齐全。
(Prepare all the ingredients.)

② 在温水中加入白醋。
(Add white vinegar into warm water.)

③ 将鸡蛋去壳后倒入水中。
(Remove the egg from the shell and pour into the water.)

④ 用勺子搅拌鸡蛋。
(Stir the eggs with a spoon.)

⑤ 水波蛋煮熟后捞起整形。
(When the eggs are cooked, remove and set.)

⑥ 最后装盘即可。
(Place the garnish and serve with the dish.)

❶

❷

❸

❹

❺

❻

带壳溏心蛋
Soft Boiled Egg with Shell

菜肴特点
(Dish Characteristic)

鲜嫩味美
Fresh and tasty

带壳溏心蛋

原 料 (Ingredient)

| 鸡蛋 | Egg | 2个 |

温馨提示 (Kindly Reminder)

选用新鲜鸡蛋。
(Choose fresh eggs.)

制作过程 (Method)

① 将所有原料准备齐全。
(Prepare all the ingredients.)

② 将低温机温度控制在65度。
(Control the temperature of 65 degree of low temperature machine.)

③ 将鸡蛋放入65度的水中煮1小时。
(Add eggs into 65 degree water pot to boil the eggs for an hour.)

④ 将煮好的鸡蛋放入凉水中凉透。
(Put the soft boiled eggs in cold water and cool thoroughly.)

⑤ 用专业开蛋器将鸡蛋壳打开。
(Use a professional egg opener to open the shell.)

⑥ 最后装盘即可。
(Place the garnish and ready to serve.)

❶

❷

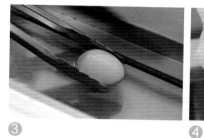
❸

❹

❺

❻

蛋黄酱水煮蛋
Poach Egg with Mayonnaise

菜肴特点
(Dish Characteristic)

色泽鲜艳,口感嫩滑
Colorful, tender, smooth creamy egg

原料 (Ingredients)

鸡蛋	Egg	150g
白醋	White vinegar	15g
蛋黄酱	Mayonnaise	20g
细葱	Chive	5g
小番茄	Cherry tomato	10g
黑橄榄	Black olives	100g
盐	Salt	3g
胡椒	Pepper	3g

温馨提示 (Kindly Reminder)

（1）鸡蛋必须选用新鲜的。
(Eggs must be fresh.)

（2）鸡蛋不要煮过头。
(Don't overcook eggs when boiling them.)

制作过程 (Method)

① 将所有原料准备齐全。
(Prepare all the ingredients.)

② 将鸡蛋放入加过白醋的温水中煮10分钟。
(Boil the eggs in warm water with white vinegar for 10 minutes.)

③ 将煮熟的鸡蛋对切开，然后取出蛋黄。
(Halved the boiled eggs and remove the egg yolk.)

④ 将蛋黄和蛋黄酱、盐、胡椒搅打均匀。
(Mix egg yolk with mayonnaise, salt and pepper.)

⑤ 将蛋黄酱泥挤入蛋白中。
(Gently squeezes the mixture of egg yolk and mayonnaise into the egg white.)

⑥ 最后加入小番茄、黑橄榄、细葱装盘即可。
(Garnish with cherry tomato, black olives chive and serve.)

❶

❷

❸

❹

❺

❻

法式番茄炖蛋
French Style Stew Egg with Tomato

菜肴特点
(Dish Characteristic)

酸甜可口，营养丰富
Sweet and sour taste, rich nutrition

鸡蛋中的蛋白质属于完全蛋白质，利于人体吸收。西红柿中的番茄红素具有独特的抗氧化作用，可以有效清除人体内的自由基，达到抗衰老、增强免疫力的作用。

Egg is an excellent source of protein which is beneficial to human absorption , while lycopene in tomatoes has a unique antioxidant effect that can effectively scavenge free radicals in the body to achieve anti-aging and enhance immunity.

法式番茄炖蛋

原料 (Ingredients)

鸡蛋	Egg	50g
番茄	Tomato	30g
黄油	Butter	15g
干葱	Shallot	10g
盐	Salt	3g
胡椒	Pepper	3g

温馨提示 (Kindly Reminder)

此菜营养丰富，方便易做。
(This dish is nutritious and easy to cook.)

制作过程 (Method)

① 将所有原料准备齐全。
(Prepare all the ingredients.)

② 将番茄切成1cm的丁，干葱切末。
(Cut the tomatoes into 1cm cubes and chop shallots.)

③ 在锅中放入黄油、干葱、番茄丁炒软，最后加入盐和胡椒调味。
(In a saucepan, add chopped butter, shallot, tomato cubes, and stir until tender. Season with salt and pepper.)

④ 将炒好的番茄酱放入碗的底部。
(Put the tomato sauce into the bottom of the bowl.)

⑤ 在番茄酱上面加入鸡蛋，放入烤箱隔水蒸20分钟。
(Add eggs on top to the tomato sauce and steam in the oven for 20 minutes.)

⑥ 最后加入装饰物装盘即可。
(Add garnishes and serve in a dish.)

第4章

小食

俱乐部三明治
Club Sandwiches

菜肴特点
（Dish Characteristic）

营养丰富，美味可口
Rich in nutrition and delicious

三明治有一个非常有趣的故事。英国一个叫 Sandwich 的小镇里有一个长期沉迷于纸牌游戏的人，为了方便边打牌边吃三餐，他就把几种食物跟面包夹在一起食用。

An interesting story about sandwiches came from a small town named Sandwich in British. A card game addict who assembles different kind of foods with bread for convenience while he was playing cards and eating his daily meals.

 俱乐部三明治

原料 (Ingredients)

吐司	Toast	3片
黄油（咸）	Butter(salty)	5g
培根	Bacon	10g
鸡肉	Chicken breast	50g
鸡蛋	Egg	1个
芝士片	Cheese slices	10g
蛋黄酱	Mayonnaise	10g
番茄	Tomato	10g
生菜	Lettuce	10g

温馨提示 (Kindly Reminder)

鸡蛋的成熟度依客人要求所定。
（How well the egg is done depends on the requirements of customers.）

制作过程 (Method)

① 将所有原料准备齐全。
（Prepare all the ingredients.）

② 将吐司表面涂上黄油，放入烤箱烤至表面金黄。
（Spread the toast with butter and bake in the oven until golden brown.）

③ 将鸡肉用基本味腌制后煎上色，然后放入烤箱烤熟待用。
（Marinate the chicken with the basic flavor, pan-fry it, then bake it until cooked.）

④ 将鸡蛋双面煎，培根煎熟，番茄切片，鸡肉切片。
（Fry eggs on both side, pan-fry bacon, slice tomato and chicken.）

⑤ 在吐司表面涂上蛋黄酱，然后和芝士片、生菜、番茄叠在一起。
（Spread the mayonnaise on sandwich and place in cheese slices, lettuce and tomato.）

⑥ 最后将三明治切成小份，配上薯条装盘即可。
（Finally, cut the sandwich into small pieces and serve it with french fries.）

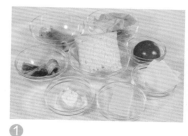
①

②

③

④

⑤

⑥

金枪鱼帕尼尼
Tuna Pan Ni Ni

菜肴特点
（Dish Characteristic）

营养丰富，简单方便
Rich in nutrition, simple and convenience

# 金枪鱼帕尼尼

原料 (Ingredients)

罐装金枪鱼	Canned tuna	150g
帕尼尼三明治	Pan Ni Ni	50g
牛油果	Avocado	50g
茄子	Eggplant	10g
节瓜	Zucchini	15g
洋葱	Onion	10g
生菜	Lettuce	10g
蛋黄酱	Mayonnaise	50g
芝士	Cheese	20g
欧芹	Parsley	5g
盐	Salt	3g
胡椒	Pepper	3g

 温馨提示 (Kindly Reminder)

可选用时令蔬菜。
(Seasonal vegetables can be used.)

制作过程 (Method)

① 将所有原料准备齐全。
(Prepare all the ingredients.)

② 将茄子、节瓜切片，然后放入热煎锅中煎熟。
(Slice and pan-fry the eggplants and zucchinis.)

③ 将牛油果切片。
(Slice the avocado.)

④ 在灌装金枪鱼中放入洋葱碎、欧芹碎、蛋黄酱、盐、胡椒拌匀。
(Add chopped onion, chopped parsley, mayonnaise, salt and pepper into the canned tuna, then mix well.)

⑤ 将拌好的金枪鱼酱涂在帕尼尼三明治上。
(Spread the tuna paste over the Pan Ni Ni.)

⑥ 最后放入生菜、芝士，切小份装盘即可。
(Put the lettuce, cheese on the Pan Ni Ni, cut into small pieces, then put them on the plate.)

❶

❷

❸

❹

❺

❻

牛肉汉堡
Beef Hamburger

菜肴特点
（Dish Characteristic）

营养丰富，鲜嫩可口
Nutritious and tasty, fresh and tender

 牛肉汉堡

原料 (Ingredients)

汉堡	Hamburger	30g
牛肉	Beef	100g
酸黄瓜	Gherkin	15g
番茄	Tomato	10g
生菜	Lettuce	10g
芝士	Cheese	10g
盐	Salt	4g
胡椒	Pepper	5g
百里香	Thyme	5g
洋葱	Onion	15g
鸡蛋	Egg	60g

温馨提示 (Kindly Reminder)

汉堡里面也可以加入鸡蛋、黄瓜等原料。
（Egg and cucumber can also be added into the hamburger.）

制作过程 (Method)

① 将所有原料准备齐全。
（Prepare all the ingredients.）

② 将酸黄瓜、番茄、洋葱切片。
（Slice the gherkin, tomato and onion.）

③ 将牛肉切末，加入洋葱末、百里香、盐、胡椒、蛋液拌匀。
（Chop the beef and add the chopped onion, thyme, salt and pepper. Mix well.）

④ 将牛肉煎饼，然后烤熟即可。
（Pan-fry the beef platter, then put in the oven and bake it until it is cooked.）

⑤ 将汉堡烤热。
（Toast the hamburger.）

⑥ 最后加入芝士片、牛肉、酸黄瓜、番茄、生菜装盘即可。
（Add cheese, beef, gherkin, tomato and lettuce, then it is ready to serve.）

❶

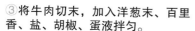
❷

❸

❹

❺

❻

热狗包配薯条佐热狗酱

Hot Dog and French Fried

菜肴特点
(Dish Characteristic)

色泽鲜艳,营养丰富
Brightly colourful and nutritious food

热狗包配薯条佐热狗酱

原料 (Ingredients)

热狗包	Hot dog	50g
香肠	Sausage	20g
柠檬	Lemon	10g
洋葱	Onion	15g
生菜	Lettuce	15g
芥末	Mustard	5g
酸黄瓜	Gherkin	10g
蛋黄酱	Mayonnaise	20g
薯条	French fries	30g

温馨提示 (Kindly Reminder)

热狗酱也可以选择番茄酱和黄芥末。
(Hot dog can also serve with tomato sauce and mustard.)

制作过程 (Method)

① 将所有原料准备齐全。
(Prepare all the ingredients.)

② 将香肠放进烤炉烤熟。
(Place the sausage in the oven and toast it .)

③ 将蛋黄酱、柠檬、芥末、洋葱碎、酸黄瓜碎搅拌均匀，制成酱汁。
(Mix mayonnaise, lemon juice, mustard, chopped onion, chopped gherkin to form the sauce.)

④ 将薯条炸至金黄色。
(Deep fry the fries until golden brown.)

⑤ 将热狗包改刀，中间涂上酱汁。
(Cut the hot dog into 1/2 and squeeze the sauce in the hot dog bun.)

⑥ 最后在热狗中加入生菜、酸黄瓜、香肠及酱汁，装盘即可。
(Place the hot dog with lettuce, gherkin, sausage and sauce, it is ready to serve.)

❶

❷

❸

❹

❺

❻

炸鱼柳配太太沙司
Fried Fish Fillet with Tatar Sauce

菜肴特点
(Dish Characteristic)

色泽鲜艳，外脆里嫩
Brightly colour, crisp outside and soft inside

原料 (Ingredients)

龙利鱼	Sole fish	200g
天妇罗粉	Flour	100g
鸡蛋	Egg	50g
柠檬	Lemon	1个
洋葱	Onion	10g
水瓜柳	Capers	10g
酸黄瓜	Gherkin	10g
蛋黄酱	Mayonnaise	50g

温馨提示 (Kindly Reminder)

也可以选用鲷鱼、鳕鱼等鱼类来炸制鱼柳。
(You may use snapper, cod and other kind fish for substitution.)

制作过程 (Method)

① 将所有原料准备齐全。
(Prepare all the ingredients.)

② 将龙利鱼切条。
(Slice the sole fish into strips.)

③ 将天妇罗粉、鸡蛋、水调成面糊。
(Mix the Flour, egg, water in a medium bowl until well combined.)

④ 将鱼柳裹上面糊，放入150℃油温锅中炸至金黄色。
(Dip the fillets in the batter and fry them at a temperature of 150℃ until it turned golden brown.)

⑤ 将洋葱碎、水瓜柳碎、酸黄瓜碎、蛋黄酱、柠檬汁搅拌成酱汁。
(Add chopped onion, chopped capers, chopped gherkin, mayonnaise and lemon juice into mixing bowl and mix well.)

⑥ 最后加入装饰物装盘即可。
(Place the garnish, then it is ready to serve.)

①

②

③

④

⑤

⑥

可丽饼
Crepe

菜肴特点
(Dish Characteristic)

色泽鲜艳，香甜软嫩
Bright colour, sweet and fluffy

可丽饼是一款用小麦粉制作的煎饼，风靡欧洲乃至世界的各个角落。传统的可丽饼是用柴火先把陶土制成的圆盘烤热，再把面糊薄薄地摊在上面煎制。

The crepe is made from wheat flour, which is popular in Europe and everywhere in the world. The traditional crepe was made with a tray made by clay and heating it up with firewood, then pan-fry it by spreading the batter thinly on it.

原料 (Ingredients)

鸡蛋	Egg	50g
奶油	Cream	200g
糖	Sugar	5g
芒果	Mango	100g
牛奶	Milk	200g
蓝莓	Blueberry	10g
面粉	Flour	100g
草莓	Strawberry	10g
红加仑	Redcurrant	10g

温馨提示 (Kindly Reminder)

可丽饼可用于早餐，原味食用也可以。
(Crepe can be served for breakfast, or just serve it when it is ready.)

制作过程 (Method)

① 将所有原料准备齐全。
(Prepare all the ingredients.)

② 将鸡蛋、糖、牛奶、面粉混合均匀后倒入不粘锅。
(Mix the eggs, sugar, milk and flour well, add batter and swirl completely and cover the bottom of skillet.)

③ 将其煎成金黄色后起锅，冷却后用模具弄成小圆饼。
(Cook until underside of crepe is golden brown, mold a small round cake after cooling.)

④ 将芒果切丁。
(Cut mango into small dice.)

⑤ 将奶油打发后挤在可丽饼上。
(Whisk the cream and squeeze it on to crepes layer.)

⑥ 最后放入蓝莓、草莓、红加仑等水果装盘即可。
(Add blueberry, strawberry, redcurrant and other fruits for decoration, then it is ready to serve.)

①

②

③

④

⑤

⑥

橄榄面包
Black Olive Toast

菜肴特点
(Dish Characteristic)

口感松脆,奶香可口
Sliced baguette topped with olive, it tastes crispy and creamy

橄榄面包

原料 (Ingredients)

法棍	Baguette	100g
黄油	Butter	20g
黑橄榄	Blake olive	20g
欧芹	Parsley	10g
盐	Salt	5g
胡椒	Pepper	4g

温馨提示 (Kindly Reminder)

盐和胡椒可以放也可以不放，按个人口味定。
(You may season with salt and pepper.)

制作过程 (Method)

① 将所有原料准备齐全。
(Prepare all ingredients.)

② 将法棍切片。
(Slice the baguette.)

③ 将黑橄榄切末。
(Chop the black olives.)

④ 在法棍面包片上涂上黄油，撒上黑橄榄碎，放入烤箱用180℃烤10分钟。
(Spread the baguette with butter, add chopped black olives on top, then bake at 180℃ for 10 minutes.)

⑤ 在烤好的面包片上撒上欧芹碎、盐和胡椒。
(Sprinkle the chopped parsley, salt and pepper on the baked baguettes.)

⑥ 最后装盘即可。
(Transfer to a serving platter, then it is ready to serve.)

❶

❷

❸

❹

❺

❻

橙味法式芝多士
French Toast with Orange Sauce

菜肴特点
(Dish Characteristic)

色泽鲜艳，外脆里嫩
Such a colorful dish, perfectly crisp outside and soft inside

原料 (Ingredients)

吐司	Toast	200g
奶油	Cream	100g
牛奶	Milk	100g
鸡蛋	Egg	150g
蓝莓	Blueberry	10g
红加仑	Cranberry	10g
橙子	Orange	150g
蜂蜜	Honey	50g
黄油	Butter	10g
糖	Sugar	8g

温馨提示 (Kindly Reminder)

吐司可以不加橙汁，做成其他口味或原味都可以。
(Toast can be added to other flavors instead of orange juice.)

制作过程 (Method)

① 将所有原料准备齐全。
(Prepare all the ingredients.)

② 将橙子取肉留汁。
(Cut the orange into wedges and keep the orange juice.)

③ 将牛奶、鸡蛋、橙汁、糖、蜂蜜搅拌均匀后放入吐司浸泡。
(Dunk each slice of toast into the mixture of milk, egg, orange juice, sugar and honey, soaking both sides.)

④ 将浸泡好的吐司放在热黄油锅上煎至金黄色。
(Add butter to the pan, place the soaked toast on pan, and cook on both sides until golden. Repeat with additional pieces.)

⑤ 将吐司改刀成型。
(Slice the toast into triangle shape.)

⑥ 挤上打发好的奶油，放上红加仑和蓝莓装盘即可。
(Top with whipped cream, blueberry, cranberry and it is ready to serve.)

❶

❷

❸

❹

❺

❻

香酥鸡米花
Popcorn Chicken

菜肴特点
(Dish Characteristic)

色泽鲜艳,外脆里嫩,肉质鲜美
Bright colour, crispy and tender outside, tasty

原料 (Ingredients)

鸡腿肉	Chicken thigh	200g
面粉	Flour	100g
鸡蛋	Egg	100g
面包糠	Breadcrumb	100g
辣椒	Chili	5g
咖喱粉	Curry powder	5g
番茄酱	Tomato sauce	10g
盐	Salt	3g
胡椒	Pepper	3g

温馨提示 (Kindly Reminder)

控制好油温。
(Maintaining proper oil temperature.)

制作过程 (Method)

① 将所有原料准备齐全。
（Prepare all the ingredients.）

② 将鸡腿肉切成小块。
（Dice the chicken thigh.）

③ 将鸡腿肉用盐、胡椒、辣椒末、咖喱粉调味。
(Marinate the chicken thigh with chopped chili, season with salt, pepper and curry powder.)

④ 将腌好的鸡肉裹上面粉、鸡蛋液、面包糠。
(Add the chicken with the flour and egg mixture. Next transfer the chicken into the breadcrumb. Mix well.)

⑤ 将处理好的鸡肉放入油锅中炸至金黄色。
(Heat the oil, deep fry the chicken evenly till golden.)

⑥ 最后加上装饰物装盘即可，可蘸番茄酱吃。
(Place the garnish, then it is ready to serve, you can take it with tomato sauce.)

❶

❷

❸

❹

❺

❻

第5章

汤类

西班牙冷汤
Gazpacho

菜肴特点
(Dish Characteristic)

营养丰富，酸爽可口
Nutritious, sweet and sour

西班牙冷汤

原 料 (Ingredients)

番茄	Tomato	150g
洋葱	Onion	30g
面包	Bread	30g
红酒醋	Red wine vinegar	5g
彩椒	Capsicum	50g
小番茄	Cherry tomato	50g
黄瓜	Cucumber	30g
橄榄油	Olive oil	10g
盐	Salt	5g
胡椒	Pepper	5g

温馨提示 (Kindly Reminder)

这道菜以番茄为基底，是一道很好的开胃菜，适合夏季食用。

(A great gazpacho recipe starts with epic tomatoes. This is a truly beautiful soup for serving ice-cold during the summer and it's particularly refreshing.)

制作过程 (Method)

① 将所有原料准备齐全并且清洗干净。
(Prepare all the ingredient and clean them thoroughly.)

② 将所有蔬菜（番茄、洋葱、彩椒、小番茄、黄瓜）加工成块，面包去边切丁备用。
(Dice all the vegetables including tomato, onion, capsicum, cherry tomato, cucumber, cut off the bread crust and cut the remainder into dice. Set aside.)

③ 将所有蔬菜和面包丁放入搅拌机中。
(Put all the vegetables and croutons in the blender.)

④ 在搅拌机中加入橄榄油、红酒醋、盐和胡椒，将原料打成浓汤汁。
(Add olive oil, red wine vinegar, salt and pepper then blend into a thick soup.)

⑤ 将打好的汤放入冰箱冷藏30分钟，然后倒入盘中。
(Freeze the thick soup in the fridge for 30 minutes, then pour in the bowl.)

⑥ 装盘时加入装饰物，淋上橄榄油即可。
(Garnish and pour with olive oil, then serve.)

①

②

③

④

⑤

⑥

意大利蔬菜汤
Italian Minestrone Soup

菜肴特点
（Dish Characteristic）

酸甜可口，营养丰富
Sweet and sour, rich in nutrition

原料 (Ingredients)

土豆	Potato	200g
洋葱	Onion	20g
番茄	Tomato	30g
胡萝卜	Carrot	20g
茄膏	Tomato paste	25g
圆白菜	Cabbage	15g
欧芹	Parsley	10g
大葱	Leek	10g
大蒜	Garlic	5g
意大利面	Pasta	10g
芝士粉	Cheese powder	10g
盐	Salt	5g
胡椒	Pepper	4g
西芹	Celery	20g

温馨提示 (Kindly Reminder)

可以根据季节选用时令蔬菜。
（Other seasonal vegetables can be used for this soup.）

制作过程 (Method):

① 将所有原料准备齐全。
（Prepare all the ingredients.）

② 将土豆、洋葱、番茄、西芹、胡萝卜、圆白菜、大葱、大蒜都切成1厘米的丁。
（Cut potato, onion, tomato, celery, carrot, cabbage, leek and garlic into 1cm cubes.）

③ 在锅中放入蔬菜丁炒香。
(Stir-fry the vegetables.)

④ 蔬菜炒香后加入茄膏，加入水煮20分钟，然后加入意大利面，用盐和胡椒调味。
(Add the tomato paste, simmering for 20 minutes. Add pasta and season with salt and pepper.)

⑤ 将蔬菜汤倒入盘中。
(Pour the vegetable soup into the plate.)

⑥ 最后加入芝士粉、欧芹，装盘即可。
(Garnish with cheese powder and parsley, then serve.)

奶油蘑菇汤
Creamy Mushroom Soup

菜肴特点
(Dish Characteristic)

奶香浓郁，咸鲜可口，营养丰富
Rich in milk, salty and delicious, rich in nutrition

　　奶油蘑菇汤是一道经典的法国菜，以咸鲜为主，其中的营养成分不可忽视。首先，蘑菇可以增强免疫力。其次，洋葱有杀菌、祛风寒的作用，在寒冷的冬天可以带来如棉袄般的温暖。
　　Cream mushroom soup is a classic French dish, mainly salty and fresh.But the nutritional components should not be ignored.Firstly,mushrooms can enhance immune system. Secondly,onion also has the function of sterilization ,which can prevent from getting sick. In cold winter, you can feel the warmth of a cotton padded jacket when you drink it.

原料 (Ingredients)

蘑菇	Mushroom	50g
奶油	Cream	100g
牛奶	Milk	30g
面粉	Flour	10g
黄油	Butter	10g
干葱	Shallot	10g
盐	Salt	3g
胡椒	Pepper	3g

温馨提示 (Kindly Reminder)

此菜营养丰富，方便易做。
(This dish is nutritious and easy to cook.)

制作过程 (Method)

① 将所有原料准备齐全。
(Prepare all the ingredients.)

② 将蘑菇切成1cm的丁，干葱切末。
(Cut the mushroom into 1cm cubes and finely chopped shallots.)

③ 在锅中放入黄油、干葱、蘑菇丁炒香，然后加入牛奶、奶油、面粉，用盐和胡椒调味。
(Stir-fry chopped shallot, mushroom cubes, then add in milk, cream, flour. Season with salt and pepper.)

④ 将炒好的蘑菇汤放入搅拌机中打成汤汁。
(Pour the mushroom soup into the blender.)

⑤ 将奶油蘑菇汤装入汤盘。
(Pour the cream mushroom soup into the soup bowl.)

⑥ 最后加入蘑菇片装盘即可。
(Garnish with slice cooked mushroom, then it is ready to serve.)

①

②

③

④

⑤

⑥

马赛鱼汤
Bouillabaisse

菜肴特点
(Dish Characteristic)

味道鲜美,营养丰富
It is very tasty and full of nutrition

马赛鱼汤

原料 (Ingredients)

鱼肉	Fish meat	50g
鱿鱼	Squid	5g
虾	Prawn	20g
洋葱	Onion	10g
贝壳	Shell	20g
藏红花	Saffron	5g
青口	Mussel	10g
干白	White wine	5g
大蒜	Garlic	5g
番茄酱	Tomato sauce	10g
番茄	Tomato	10g
盐	Salt	5g
胡椒	Pepper	4g

温馨提示 (Kindly Reminder)

马赛鱼汤来自巴黎。
(Bouillabaisse is from Paris.)

制作过程 (Method)

① 将所有原料准备齐全。
(Prepare all ingredients.)

② 将鱼肉切块。
(Cut the fish into pieces.)

③ 在炒锅中加入洋葱末、大蒜末炒香，再加入鱿鱼、虾、贝壳、青口和干白炒香。
(Pan-fry the onion and garlic, then add in squid, prawn, shell, mussel and white wine.)

④ 在锅里加入藏红花汤汁、番茄、番茄酱、盐、胡椒，煮15分钟。
(Add the saffron soup, tomato, tomato sauce, salt and pepper, then simmer for 15 minutes.)

⑤ 将鱼肉煎上色后装在盘子底部。
(Pan-fry the fish and place it on the bottom of the plate.)

⑥ 倒入海鲜汤，加上装饰物，装盘即可。
(Pour the soup and garnish, then it is ready to serve.)

❶

❷

❸

❹

❺

❻

罗宋汤
Borsch

菜肴特点
（Dish Characteristic）

营养丰富，酸甜带辣
Rich in nutrition,sweet,sour and spicy

罗宋汤

原料 (Ingredients)

牛肉	Beef	150g
黄油	Butter	10g
圆白菜	Cabbage	10g
小米椒	Chili	5g
酸奶油	Sour cream	5g
百里香	Thyme	5g
红菜头	Beetroot	30g
洋葱	Onion	10g
牛清汤	Beef broth	300g
胡萝卜	Carrot	10g
番茄	Tomato	30g
欧芹	Parsley	5g
大蒜	Garlic	10g
香叶	Bay leave	5g
茄膏	Tomato paste	50g
土豆	Potato	10g
盐	Salt	5g
胡椒	Pepper	4g

 温馨提示 (Kindly Reminder)

罗宋汤里的圆白菜在快起锅之前放入，这样圆白菜可以保持脆爽。
(Simmering the cabbage right before the soup is done, this step is to remain the crispy of the cabbage .)

制作过程 (Method)

① 将所有原料准备齐全。
(Prepare all the ingredients.)

② 将牛肉切丁。
(Dice the beef in cubes.)

③ 将土豆、番茄、洋葱、红菜头、小米椒、圆白菜、胡萝卜等蔬菜切丁。
(Dice potatoes, tomatoes, onions, beetroots, chilis, cabbage, carrot and other vegetables in cubes.)

④ 将所有原料加工成丁备用。
(Dice every ingredient in cubes.)

⑤ 在锅中加入黄油、牛肉和蔬菜炒香，加入茄膏、百里香、牛清汤、大蒜、香叶，然后用盐、胡椒调味。
(Stir-fry beef and vegetable, add in tomato paste, thyme, beef broth, garlic, bay leave, and season with salt and pepper.)

⑥ 最后装盘时加入酸奶油和欧芹即可。
(Garnish with sour cream and parsley, then serve.)

❶

❷

❸

❹

❺

❻

鲜虾浓汤
Shrimp Bisque

菜肴特点
(Dish Characteristic)

口感鲜美,营养丰富
Nourishing and delicious

鲜虾浓汤

原 料 (Ingredients)

虾	Prawn	100g
西芹	Celery	10g
虾干	Dried shrimp	20g
百里香	Herb	5g
洋葱	Onion	20g
香叶	Bay leave	5g
胡萝卜	Carrot	10g
茄膏	Tomato paste	5g
盐	Salt	5g
奶油	Cream	5g
胡椒	Pepper	4g

 温馨提示 (Kindly Reminder)

龙虾浓汤也可以用这种做法。
(This recipe can be used for making lobster bisque.)

制作过程 (Method)

① 将所有原料准备齐全。
(Prepare all ingredients.)

② 将虾和虾干放烤箱烤干。
(Place dried shrimps and prawn on the baking try and put it in the oven until it dries out.)

③ 将洋葱、胡萝卜、西芹切丁。
(Dice the onions, carrots and celery in cubes.)

④ 将蔬菜、香料炒香，然后放入茄膏、虾干、虾汤、香料（香叶、百里香），用盐和胡椒调味，加入水煮15分钟。
(Stir-fry the vegetables and spices, add tomato paste, dried shrimp,shrimp soup and spice herb,bay leave, seasoning with salt and pepper,then simmer for about 15 minutes.)

⑤ 将汤搅打成汁，然后过滤。
(Blend the mixture into soup, strain it in a sieve.)

⑥ 最后放入奶油搅拌入味后装盘即可。
(In the end, stir in cream until just thickened. Then it is ready to serve.)

①

②

③

④

⑤

⑥

第6章

配 菜

芝士舒芙蕾
Cheese Soufflé

菜肴特点
(Dish Characteristic)

口感软嫩，奶香可口
Soft, tender and creamy

原料 (Ingredient)

格律耶尔芝士	Gruyere Cheese	100g
帕玛森芝士粉	Parmesan Cheese Powder	10g
白葡萄酒	White wine	10g
鸡蛋	Egg	20g
奶油	Cream	20g

温馨提示 (Kindly Reminder)

根据个人口味,盐和胡椒可以放也可以不放。
(Amount of salt and pepper can be adjusted according to personal taste.)

制作过程 (Method)

① 将所有原料准备齐全。
(Prepare all the ingredients.)

② 将蛋黄加白葡萄酒隔热打成泡沫状。
(Beat the egg yolks with white wine until frothy.)

③ 将蛋白打成鸡尾状。
(Beat the egg whites until stiff.)

④ 将格律耶尔芝士和奶油加热至融化,然后倒入碗的底部。
(Melt Gruyere cheese and cream, and pour into the bottom of the bowl.)

⑤ 在芝士和奶油上放上搅拌在一起的蛋液,撒上帕玛森芝士粉,然后放入烤箱用180℃烤10分钟至金黄色。
(Put egg mixture on top of the melt cheese and cream, sprinkle with Parmesan Cheese Powder, and bake for 10 minutes at 180℃ until golden brown.)

⑥ 最后装盘上桌即可。
(Finally serve on a plate.)

①

②

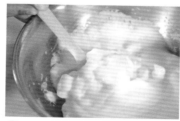
③

④

⑤

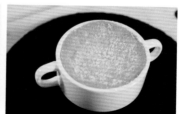
⑥

土豆泥
Mashed Potato

菜肴特点
(Dish Characteristic)

造型美观，香嫩可口
Nice in presentation, creamy, tender and delicious

土豆泥

原 料 (Ingredients)

土豆	Potato	200g
奶油	Cream	30g
牛奶	Milk	25g
黄油	Butter	10g
盐	Salt	5g
胡椒	Pepper	4g

温馨提示 (Kindly Reminder)

（1）土豆要选新鲜的，不要选发芽的、发黑的、烂的土豆。
(Choose fresh potatoes, don't pick sprouted, blackened or rotten ones.)

（2）土豆泥可搭配牛肉、鸡肉、鱼等菜肴食用。
(Mashed potatoes can also be served with beef, chicken, fish and other dishes.)

制作过程 (Method)

① 将所有原料准备齐全。
（Prepare all the ingredients.）

② 将土豆切片，放在水里煮熟。
（Slice the potatoes and boil them in the water.）

③ 将煮好的土豆挤成泥。
(Mash the boiled potatoes.)

④ 在锅中加入土豆泥、奶油、牛奶、黄油、盐和胡椒，搅拌成团。
(Add in mashed potatoes, cream, milk, butter, salt and pepper to form a smooth dough.)

⑤ 将土豆泥挤入盘中。
(Squeezes the mashed potatoes into the plate.)

⑥ 最后加入装饰物即可。
(Finally place decorations and ready to serve.)

❶

❷

❸

❹

❺

❻

蒜香意面
Garlic Pasta

菜肴特点
（Dish Characteristic）

营养丰富，口感软嫩
Rich in nutrition, soft and tender taste

蒜香意面

原料 (Ingredients)

意大利面	Pasta	200g
红椒粉	Chili pepper	10g
黄油	Butter	10g
干葱	Shallot	20g
大蒜	Garlic	10g
橄榄油	Olive oil	10g
芝士粉	Cheese powder	5g
盐	Salt	5g
胡椒	Pepper	4g

温馨提示 (Kindly Reminder)

意大利面焯水不要焯过头。
（Blanch pasta by boiling water, do not overcook.）

制作过程 (Method)

① 将所有原料准备齐全。
（Prepare all the ingredients.）

② 将意面放入沸水中煮熟。
（Cook the spaghetti in boiling water.）

③ 意面焯好后放入橄榄油。
（Blanch the spaghetti and add the olive oil.）

④ 将大蒜切片。
（Slice the garlic.）

⑤ 在炒锅中加入黄油、大蒜片、红椒粉、盐、胡椒炒香，然后放入意大利面。
（Add butter, garlic flakes, chili pepper, salt and pepper to the pan, and then pasta.）

⑥ 最后撒上芝士粉，加入干葱，装盘即可。
（Sprinkle with cheese powder, shallot and serve.）

❶

❷

❸

❹

❺

❻

洋葱培根薯角

Onion and Bacon with Potato Wedges

菜肴特点
（Dish Characteristic）

外脆里嫩，美味可口
Crispy outside, tender inside, delicious

洋葱培根薯角

原料 (Ingredients)

土豆	Potato	250g
培根	Bacon	10g
洋葱	Onion	100g
欧芹	Parsley	10g
胡椒	Pepper	4g
盐	Salt	5g

温馨提示 (Kindly Reminder)

炸土豆的时候土豆角外面可以裹一些面粉，防止土豆水分过多导致安全事故。
(When deep-frying potatoes, the potato wedges can be wrapped with some flour to prevent excessive moisture of potatoes. Otherwise, safety issue might occur.)

制作过程 (Method)

① 将所有原料准备齐全。
(Prepare all the ingredients.)

② 将土豆切成土豆角。
(Cut the potatoes into potato wedges.)

③ 将土豆角放入150℃油锅中炸至金黄色。
(Put potato wedges in 150℃ oil until golden brown.)

④ 将洋葱、培根、欧芹切末。
(Chop the shallots, bacon and parsley.)

⑤ 将洋葱末、培根末放在热锅中炒香，然后放入炸好的土豆角、盐、胡椒翻炒均匀。
(Stir-fry the chopped spring onion and bacon in a hot pan. Add the wedges, salt and pepper. Stir well.)

⑥ 最后撒上欧芹末装盘即可。
(Finally sprinkle with chopped parsley and serve on a plate.)

香草烤杂蔬
Roasted Herb Mixed Vegetable

菜肴特点
（Dish Characteristic）

营养丰富，简单易做，咸鲜可口
Rich nutrition, easy to cook, salty delicious

香草烤杂蔬

原料 (Ingredients)

土豆	Potato	50g
小番茄	Cherry tomato	30g
彩椒	Capsicum	30g
茄子	Eggplant	30g
洋葱	Onion	20g
胡萝卜	Carrot	35g
大蒜	Garlic	10g
花菜	Cauliflower	40g
胡椒	Pepper	5g
橄榄油	Olive oil	20g
红薯	Sweet potato	30g
盐	Salt	5g
百里香	Thyme	5g

温馨提示 (Kindly Reminder)

蔬菜可选用各种时令蔬菜。
(Vegetables can be selected from various seasonal vegetables.)

制作过程 (Method)

① 将所有原料准备齐全并清洗干净。
(Prepare and clean all the ingredients.)

② 将所有的蔬菜都切配成型。
(Cut and shape all the vegetables.)

③ 用橄榄油、盐、胡椒、百里香腌制蔬菜（土豆、小番茄、彩椒、茄子、洋葱、胡萝卜、大蒜、花菜、红薯）。
(Marinate vegetables including potato,cherry tomato,capsicum,eggplant, onion, carrot,garlic,cauliflower,and sweet potato with olive oils, salt, pepper and thyme.)

④ 将腌制好的蔬菜放入180℃的烤箱烤30分钟。
(Place the marinated vegetables in a 180℃ pre-heated oven and bake for 30 minutes.)

⑤ 待蔬菜烤熟后从烤箱取出。
(Remove the vegetables from the oven when it turns brown and cooked.)

⑥ 最后装盘即可。
(Ready to serve.)

①

②

③

④

⑤

⑥

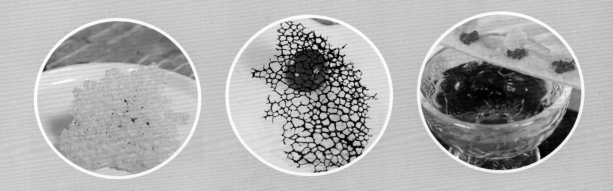

第7章 装饰配菜

墨鱼薄脆
Cuttlefish Chip

菜肴特点
(Dish Characteristic)

造型美观，可用于菜肴装饰
Nice appearance, you may use it for dish decoration

墨鱼薄脆

(Ingredients)

墨鱼汁	Cuttlefish sauce	10g
面粉	Flour	10g
油	Oil	25g
水	Water	50g

温馨提示 (Kindly Reminder)

（1）煎制的时候需用小火。
(Use low heat when frying.)

（2）墨鱼薄脆可以为海鲜、羊排等菜肴做装饰。
(Chip can be decorated with seafood, lamb chops and other dishes .)

制作过程 (Method)

① 将所有原料准备齐全。
(Prepare all the ingredients.)

② 将所有原料都混合在一起，搅拌均匀。
(Mix all the ingredients together and stir well.)

③ 在热锅中倒入搅拌好的液体。
(Add the mixture in pan.)

④ 用小火慢煎。
(Try it with the low heat.)

⑤ 当所有水分都煎干后起锅。
(When all the water is dry, take out the chip.)

⑥ 最后加装饰物装盘即可。
(Add decorations and it's ready to serve.)

❶

❷

❸

❹

❺

❻

芝士薄片
Cheese Crisp

菜肴特点
(Dish Characteristic)

造型美观，简单易做，香脆可口
Beautiful shape, easy to make, crisp and delicious

# 芝士薄片

原 料 (Ingredients)

| 帕玛森芝士 | Parmesan Cheese | 100g |

 温馨提示 (Kindly Reminder)

芝士最好是现做现磨。
(Freshly ground cheese is better.)

制作过程 (Method)

① 准备一块完整的帕玛森芝士。
(Prepare a thick slice of Parmesan cheese.)

② 将帕玛森芝士块磨成粉末状。
(Grind the Parmesan Cheese into powder.)

③ 将现磨的芝士末均匀地撒在不粘垫上。
(Spread the freshly ground cheese evenly over the non-sticky mat.)

④ 将芝士末放入180℃的烤箱烤10分钟。
(Place the marinated vegetables in a 180℃ oven and bake for 10 minutes.)

⑤ 待芝士烤上色后从烤箱取出。
(Take it out when it turns brown.)

⑥ 最后加装饰物装盘即可。
(Add decorations and serve on a plate.)

❶

❷

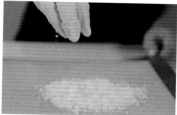
❸

❹

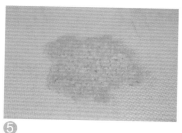
❺

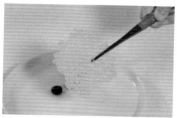
❻

菠菜芝士条
Spinach with Cheese Stick

菜肴特点
（Dish Characteristic）

口感松脆，奶香可口
Crispy and creamy

(Ingredients)

面粉	Flour	100g
芝士粉	Cheese powder	20g
黄油	Butter	20g
菠菜汁	Spinach juice	5g

温馨提示 (Kindly Reminder)

此菜可作为各种菜肴的装饰。
(It can be used for decoration of various dishes.)

制作过程 (Method)

① 将所有原料准备齐全。
(Prepare all the ingredients.)

② 将所有原料都混合在一起揉成面团。
(Mix all the ingredients together and form a dough.)

③ 将面团杆成薄片。
(Make the dough into a thin layer chip.)

④ 用刀将薄片切成型放入烤箱,用 180℃烤 10 分钟。
(Cut the sheet into an knife shape, bake it for 10 minutes at 180℃.)

⑤ 将烤好的薄片取出烤箱。
(Take the baked chip out of the oven.)

⑥ 最后加装饰物装盘即可。
(Add decorations and serve on a plate.)

❶

❸

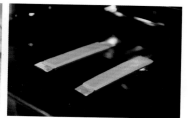
❹

❺

❻

墨鱼西米薄脆
Cuttlefish with Sago Chip

菜肴特点
（Dish Characteristic）

造型美观，香脆可口
Beautiful shape, crispy and delicious

 墨鱼西米薄脆

原料 (Ingredients)

| 墨鱼汁 | Cuttlefish sauce | 10g |
| 西米 | Sago | 100g |

温馨提示 (Kindly Reminder)

西米需要煮至透明，然后过冷水后再沥干水分。
(Properly cooked sago will be translucent, then rinse sago with cold running water and drained.)

制作过程 (Method)

① 准备好西米和墨鱼汁。
(Prepare sago and cuttlefish sauce.)

② 将西米放入沸水中煮30分钟，期间需要不停地搅拌，防止粘锅。
(Slowly add sago in boiling water until mostly translucent with constant stirring about 30 minutes.)

③ 将煮好的西米与墨鱼汁混合均匀。
(Mix the sago with the cuttlefish sauce.)

④ 将墨鱼西米均匀地放在不粘垫上，然后放入100℃的烤箱烤1个小时。
(Place the cuttlefish and sago evenly on a non-stick tray or silicone mat and bake in a 100℃ oven for an hour.)

⑤ 将烤干的西米取出后放在180℃的油温中炸定型。
(Remove the dried sago, and deep fried in a 180℃ and set the shape.)

⑥ 最后加装饰物装盘即可。
(Add decorations and it's ready to serve.)

❶

❷

❸

❹

❺

❻

第 8 章

主　菜

惠灵顿牛排
Beef Wellington

菜肴特点
（Dish Characteristic）

营养丰富，口感软嫩
Rich in nutrition, soft and tender taste

惠灵顿牛排

原料 (Ingredients)

牛柳	Tenderloin	250g
帕尔玛火腿片	Parma Ham	10g
干葱	Shallot	20g
酥皮	Puff pastry	100g
彩椒	Bell pepper	10g
橄榄油	Olive oil	10g
芥末	Mustard	5g
白蘑菇	Mushroom	50g
西蓝花	Broccoli	10g
小番茄	Cherry tomato	10g
布朗酱	Brown sauce	10g
胡椒	Pepper	4g
盐	Salt	5g
鸡蛋	Egg	20g
蘑菇酱	Mushroom sauce	10g

温馨提示 (Kindly Reminder)

酥皮最好现做现用，这样容易起酥。
(It is better to use fresh handmade puff pastry which is more crispy.)

制作过程 (Method)

① 将所有原料准备齐全。
(Prepare all the ingredients.)

② 将牛柳表面煎上色，然后涂上芥末、蘑菇酱，用帕尔玛火腿片包成卷后放进冰箱冷藏20分钟。
(Season the tenderloin generously with mustard, mushroom sauce. Wrap it with Parma Ham and freeze for 20 minutes.)

③ 将牛柳卷外面包上酥皮。
(Wrap the beef fillet rolls with puff pastry.)

④ 将酥皮牛柳卷刷上蛋液，然后放入烤箱用180℃烤20分钟。
(Egg wash the beef rolls, and bake it at 180℃ for 20 minutes.)

⑤ 锅中倒入橄榄油，将蔬菜（干葱、彩椒、白蘑菇、西蓝花、小番茄）加工成形后炒熟，加入盐、胡椒调味。
(Put the olive oil in the pan and stir - fry the vegetables including shallot,bell pepper,mushroom, broccoli ang cherry tomato, season with salt and pepper.)

⑥ 加入布朗酱点缀，装盘即可。
(Add Brown Sauce, then it is ready to serve.)

❶

❷

❸

❹

❺

❻

香煎三文鱼
Pan-fried Salmon

菜肴特点
(Dish Characteristic)

皮脆肉嫩，营养丰富
Crispy and golden outside, juicy and moist inside, nutritious

原料 (Ingredients)

三文鱼	Salmon	250g
青柠	Lime	5g
芦笋	Asparagus	50g
土豆	Potato	10g
柠檬	Lemon	10g
蟹味菇	Crab mushroom	10g
黄油	Butter	10g
小番茄	Cherry tomato	5g
盐	Salt	5g
小玉米	Finger corn	10g
胡椒	Pepper	5g
手指萝卜	Finger carrot	15g
百里香	Thyme	2g

温馨提示 (Kindly Reminder)

三文鱼可以搭配柠檬黄油汁、奶油汁等食用。
(The salmon can be served with lemon butter sauce and cream sauce.)

制作过程 (Method)

① 将所有原料准备齐全并洗净。
(Prepare all the ingredients and wash them thoroughly.)

② 将三文鱼用盐、胡椒、百里香、柠檬汁腌制。
(Marinate salmon with salt, pepper, thyme and lemon juice.)

③ 将三文鱼放在热锅中用黄油煎上色。
(Heat the pan, fry the salmon with butter until it turns golden.)

④ 将煎好的三文鱼放入烤箱烤至所需成熟度。
(Place the pan-fried salmon on a baking tray and put it in oven until it is cooked.)

⑤ 将土豆、蟹味菇、小番茄、小玉米、芦笋放入热油锅煎熟，加入盐和胡椒调味。
(Fry potato, crab mushroom, cherry tomato, finger corn, asparagus, and season with salt and pepper.)

⑥ 最后加入手指萝卜、青柠及其他装饰物装盘即可。
(Garnish with finger carrot, lime, and it is ready to serve.)

❶

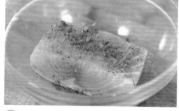
❷

❸

❹

❺

❻

意式焗龙虾
Italian Style Baked Lobster

菜肴特点
（Dish Characteristic）

口感鲜甜，奶香浓郁
Fresh and sweet, rich in milk flavor

意式焗龙虾

(Ingredients)

龙虾	Lobster	100g
帕玛森芝士粉	Parmesan Cheese Powder	20g
黄油	Butter	20g
百里香	Thyme	5g
面粉	Flour	20g
小番茄	Cherry tomato	5g
奶油	Cream	10g
柠檬	Lemon	5g
盐	Salt	5g
食锦蔬菜	Mix leaves	
胡椒	Pepper	4g
橄榄油	Olive oil	10g

温馨提示 (Kindly Reminder)

面酱 = 黄油 + 面粉。
(Roux sauce = butter + flour.)

制作过程 (Method)

①将所有原料准备齐全。
(Prepare all the ingredients.)

②将龙虾对半切开，去肠，洗净。
(Cut the lobster in half, remove the intestines and wash it clean.)

③用橄榄油将龙虾表面煎上色。
(Pan - fry the lobster with olive oil.)

④用面酱加奶油制成奶白汁，加盐、胡椒和百里香，浇在龙虾上面，并滴上几滴柠檬汁。
(Pour the roux sauce over the lobster, and add salt, pepper, thyme and lemon juice.)

⑤在龙虾上面撒上帕玛森芝士粉，进烤箱180℃烤8分钟。
(Sprinkle the lobster with Parmesan Cheese Powder and bake it at 180 ℃ for 8 minutes.)

⑥最后配上小番茄、什锦蔬菜色拉，装盘即可。
(Garnish with cherry tomato and mix leaves, and it is ready to serve.)

❶ ❷

❸ ❹

❺ ❻

油封香草鸭
Confit Duck Leg

菜肴特点
(Dish Characteristic)

肉质软嫩,口感咸鲜
Tender, salty and delicious

原料 (Ingredients)

鸭腿	Duck leg	100g
香叶	Coriander	10g
小番茄	Cherry tomato	20g
混合蔬菜	Mix leaves	20g
橄榄油	Olive oil	20g
百里香	Thyme	5g
盐	Salt	5g
胡椒	Pepper	4g
蓝莓	Blueberry	20g

温馨提示 (Kindly Reminder)

需要将鸭腿的肥肉去掉。
(You need remove the duck fat.)

制作过程 (Method)

① 将所有原料准备齐全。
(Prepare all the ingredients.)

② 将鸭腿的肥肉去掉。
(Remove the duck fat.)

③ 在橄榄油中加入百里香、盐、胡椒、香叶和鸭腿，然后放入100℃烤箱中烤2个小时。
(Add thyme, salt, pepper, bay leaves into olive oil and place the duck leg, then bake it at 100℃ oven for 2 hours.)

④ 将小番茄和蓝莓对半切开，与混合蔬菜放在一起做色拉备用。
(Cut the cherry tomatoes and blueberries, and mix vegetables for making salad.)

⑤ 将烤好的鸭腿取出。
(Take out the roasted duck leg from oven.)

⑥ 最后装盘即可。
(Garnish, then serve.)

①

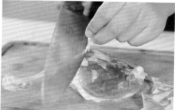
②

③

④

⑤

⑥

芝士鸡肉卷
Chicken, Ham & Cheddar Roll-ups

菜肴特点
(Dish Characteristic)

鲜嫩可口，奶香浓郁
Fresh tender, delicious and rich in milk

原 料 (Ingredients)

鸡腿	Chicken leg	250g
黄椒	Yellow pepper	10g
芝士	Cheese	30g
花菜	Cauliflower	8g
菠菜	Spinach	10g
西蓝花	Broccoli	8g
干葱	Shallot	10g
油	Oil	10g
番茄	Tomato	10g
盐	Salt	5g
胡椒	Pepper	4g

温馨提示 (Kindly Reminder)

鸡肉卷里面还可以用一些时令蔬菜来代替芝士。
(You may use seasonal vegetables instead of the cheese .)

制作过程 (Method)

① 将所有原料准备齐全。
(Prepare all the ingredients.)

② 将鸡腿肉用肉锤锤松，用盐、胡椒腌制。
(Flatten the chicken leg with meat mallet, and season with salt and pepper.)

③ 在鸡肉卷里面裹上菠菜、芝士。
(Wrap the chicken and roll it in spinach and cheese.)

④ 在热油锅中将鸡肉卷表面煎上色后放入烤箱用 180℃烤 20 分钟。
(Pan-fry chicken roll and bake for 20 minutes at 180℃ in the oven until it turns golden brown.)

⑤ 将鸡肉卷切配成型，将蔬菜（黄椒、花菜、西蓝花、干葱、番茄）炒熟。
(Cut and shape chicken rolls. Pan-fry vegetables including yellow pepper, cauliflower, broccoli, shallot and tomato.)

⑥ 放上装饰品装盘即可。
(You may start plating, then it is ready to serve.)

❶

❷

❸

❹

❺

❻

主妇鳕鱼

Cod with Hollandaise Sauce

菜肴特点
(Dish Characteristic)

鲜嫩多汁，奶香可口
Delicious and fresh, rich in milk

原料 (Ingredients)

鳕鱼	Cod fish	200g
青柠	Lime	10g
荷兰豆	Green Bean	10g
鸡蛋	Egg	50g
刁草	Dill	5g
干葱	Shallot	10g
小番茄	Cherry tomato	10g
黄油	Butter	20g
盐	Salt	5g
柠檬	Lemon	10g
胡椒	Pepper	4g

温馨提示 (Kindly Reminder)

鳕鱼也可搭配藏红花汁或柠檬黄油汁等。
(Cod fish can be served with the saffron sauce or lemon butter sauce.)

制作过程 (Method)

① 将所有原料准备齐全。
(Prepare all the ingredients.)

② 将鳕鱼用刁草、盐和胡椒腌制后煎上色。
(Marinate cod fish with dill, salt and pepper. Pan-fry again till it turns golden.)

③ 将荷兰豆、小番茄焯水备用。
(Boiling Green Beam and cherry tomatoes, drained and set aside.)

④ 用黄油、鸡蛋、柠檬隔水打成荷兰汁后浇在鳕鱼上面。
(Beat the butter, eggs and lemon until it is thick, then pour it on the cod fish.)

⑤ 将加了荷兰汁的鳕鱼放入烤箱烤5分钟。
(Place the cod fish with hollandaise sauce and bake it for 5 minutes.)

⑥ 撒上干葱配上青柠、小番茄等装饰物装盘即可。
(Garnish with shallot, lime and cherry tomatoes, then it is ready serve.)

❶

❷

❸

❹

❺

❻

黄油鸡卷
Chicken Kiev

菜肴特点
（Dish Characteristic）

热量高，外脆里嫩
High calorie, crisp outside and tender inside

原料 (Ingredients):

鸡胸肉	Chicken breast	200g
土豆	Potato	20g
蘑菇	Mushroom	10g
面粉	Flour	100g
黄油	Butter	10g
胡萝卜	Carrot	10g
西蓝花	Broccoli	10g
鸡蛋	Egg	50g
面包糠	Breadcrumb	100g
干白	White wine	10g
枕头面包	Pillow bread	20g
盐	Salt	3g
胡椒	Pepper	3g

温馨提示 (Kindly Reminder)

此菜是俄罗斯传统菜肴，适合冬季吃。
(This dish is a traditional Russian dish which is suitable for winter season.)

制作过程 (Method)

① 将所有原料准备齐全。
(Prepare all the ingredients.)

② 将鸡胸肉批成薄片，用盐、胡椒、干白腌制5分钟。
(Slice the chicken breast and marinate with salt, pepper and white wine for 5 minutes.)

③ 在鸡片中裹上黄油卷成形。
(Roll the chicken with butter.)

④ 准备好鸡蛋、面粉、面包糠和鸡肉卷备用。
(Prepare eggs, flour, breadcrumb and chicken rolls, then set aside.)

⑤ 将鸡肉卷"过三关"后放入150℃的油温中炸至金黄色。
(Dip each chicken roll in the flour, then the egg and finally the breadcrumb, repeating so each chicken roll has a double coating. To cook, heat oven to 150℃. Fry the chicken roll each side until golden.)

⑥ 加入土豆、蘑菇、胡萝卜、西蓝花、枕头面包，装盘即可。
(Garish with potato, mushroom, carrot, broccoli and pillow bread, then it is ready to serve.)

宝云苏羊排
Baoyunsu Lamb Chops

菜肴特点
（Dish Characteristic）

外脆里嫩，营养丰富
Crispy and golden outside, juicy and moist inside, nutritious

原料 (Ingredients)

羊排	Lamb chop	100g
百里香	Thyme	5g
面包糠	Breadcrumb	20g
迷迭香	Rosemary	5g
青柠	Lime	10g
柠檬	Lemon	20g
大蒜	Garlic	10g
小番茄	Cherry tomato	10g
手指萝卜	Finger carrot	5g
欧芹	Parsley	3g
盐	Salt	5g
胡椒	Pepper	4g
黄芥末	Mustard	5g

温馨提示 (Kindly Reminder)

羊肉性温热，常吃容易上火。
(Having too much lamb is not recommended, it will produce excess internal heat to human body.)

制作过程 (Method)

① 将所有原料准备齐全。
(Prepare all the ingredients.)

② 将羊排用盐、胡椒、百里香和迷迭香腌制 10 分钟。
(Marinate the lamb chops with salt, pepper, thyme and rosemary for 10 minutes.)

③ 将羊排煎熟。
(Pan-fry the lamb chops until it is cooked.)

④ 在热油锅中将大蒜末炒香，加入面包糠炒干备用。
(Stir-fry garlic and breadcrumb, set aside.)

⑤ 在煎好的羊排外面涂上黄芥末，然后裹上炒好的面包糠即可。
(Spread the fried lamb chop with mustard, then dip each chop into breadcrumb to coat well.)

⑥ 配上（小番茄、手指萝卜、欧芹、柠檬、青柠）装盘即可。
(Garnish with cherry tomato, baby carrot, parsley, lemon and lime, then it is ready to serve.)

❶

❷

❸

❹

❺

❻

法式焗蜗牛
Snails in Garlic-herb Butter

菜肴特点
(Dish Characteristic)

口感鲜美，营养丰富
Delicious, freshness, nutritious

早在罗马时期，欧洲人便开始食用蜗牛。蜗牛是非常好的蛋白质来源。
Snails were eaten in Europe as early as Roman period. Snail is an excellent source of protein.

法式焗蜗牛

原料 (Ingredients)

蜗牛	Snail	150g
李派林喼汁	Worcestershire	5g
黄油	Butter	20g
大蒜	Garlic	5g
阿里根奴	Oregano	10g
芥末	Mustard	10g
胡椒	Pepper	4g
盐	Salt	5g
白兰地	Brandy	5g
干葱	Shallot	10g
欧芹	Parsley	10g

温馨提示 (Kindly Reminder)

蜗牛不能与蟹同食，否则可导致荨麻疹。
(Snails can't be eaten with crabs, it could cause a case of the Hives.)

制作过程 (Method)

① 将所有原料准备齐全。
(Prepare all the ingredients.)

② 将干葱、大蒜切末。
(Chop shallots and garlic.)

③ 在黄油中拌入干葱末、蒜末、李派林喼汁、芥末，搓成条后冷藏备用。
(Mix butter with shallots, garlic, Worcestershire and mustard. Roll it into a long stick and freeze it.)

④ 蜗牛焯水后取出肉。
(Blanching in boiling water before take meat out of the shell.)

⑤ 在煎锅中放入干葱末、蒜末、阿里根奴、蜗牛炒香，倒入白兰地，加入盐和胡椒调味。
(Pan-fry the shallot, garlic, Oregano, snail, then add some Brandy. Season with salt and pepper.)

⑥ 将黄油放在蜗牛上，放入烤箱烤5分钟，加入欧芹等点缀装盘即可。
(Fill shells with butter mixture, bake in the oven for 5 minutes. Add parsley and serve.)

❶

❷

❸

❹

❺

❻

海鲜意面
Seafood Pasta

菜肴特点
(Dish Characteristic)

微辣，风味独特
Slightly spicy, unique flavor

原料 (Ingredients)

海虾	Sea shrimp	150g
鱿鱼	Squid	100g
小米椒	Chili	5g
桑巴酱	Samba Sauce	10g
白玉菇	White mushroom	50g
小番茄	Cherry tomato	50g
意大利面	Pasta	200g
大蒜	Garlic	10g
胡椒粉	Pepper	3g
盐	Salt	3g

温馨提示 (Kindly Reminder)

鱿鱼最好是选择新鲜的。
(Fresh squid is preferred.)

制作过程 (Method)

① 将所有原料准备齐全并洗净。
(Prepare all ingredients and wash them thoroughly.)

② 将意大利面倒入沸水中煮10分钟。
(Pour pasta into boiling water for 10 minutes.)

③ 将意大利面从水里捞出后拌入橄榄油。
(Drain cooked pasta well in a colander, then mix with some olive oil.)

④ 将大蒜切成碎末，将小番茄对半切开备用。
(Chopped the garlic, cut the cherry tomato in half, set aside.)

⑤ 在热油锅中放入大蒜末、白玉菇、小番茄炒香，然后加入海虾、鱿鱼、桑巴酱、小米椒翻炒，再加入意大利面和盐、胡椒翻炒。
(Stir-fry garlic,white mushroom and cherry tomatoes,then add in sea shrimp,squid,Samba Sauce,chili,next add in pasta,salt and pepper.)

⑥ 最后装盘即可。
(Garnish, then it is ready to serve.)

❶

❷

❸

❹

❺

❻

香草面包糠焗青口

Mussels with Garlic and Breadcrumb

菜肴特点
（Dish Characteristic）

营养丰富，口感脆嫩
Nutritious, crispy and tender

香草面包糠焗青口

原料 (Ingredients)

青口	Mussel	250g
面包糠	Breadcrumb	100g
干白	White wine	10g
橄榄油	Olive oil	10g
彩椒	Pepper	30g
芝士粉	Cheese powder	10g
洋葱	Onion	20g
黄油	Butter	6g
柠檬	Lemon	10g
迷迭香	Rosemary	5g
大蒜	Garlic	10g
盐	Salt	5g
胡椒	Pepper	5g

温馨提示 (Kindly Reminder)

青口必须选用新鲜的。
(You must choose or buy mussels alive and in the shell.)

制作过程 (Method)

① 将所有原料准备齐全。
(Prepare all the ingredients.)

② 将洋葱、彩椒、大蒜、迷迭香切小粒。
(Chop onions, peppers, garlic and rosemary into small pieces.)

③ 在热黄油锅中加入蔬菜粒和面包糠，加入盐和胡椒翻炒干，再加入少许橄榄油和芝士粉。
(Stir-fry veyetable pieces and breadcrumb with butter, then add in salt and pepper, next add in some olive oil and cheese powder.)

④ 将青口用沸水焯水，并加入干白和柠檬汁，取肉，留一半的壳。
(Boiling the mussel and adding in white wine and lemon juice, detach the meat from each shell and place it back in half shell.)

⑤ 将炒好的香草面包糠撒在青口上，放进烤箱烤5分钟。
(Place the fried breadcrumb put on the mussel for 5 minutes in the oven.)

⑥ 最后装盘即可。
(Place the garnish, and ready to serve.)

❶

❷

❸

❹

❺

❻

蟹肉土豆饼

Italian Potato Cake with Crab Meat

菜肴特点
(Dish Characteristic)

口感软嫩，奶香浓郁
Crisp and creamy, rich in milk

蟹肉土豆饼

原料 (Ingredients)

蟹	Crab	100g
干葱	Shallot	10g
黄油	Butter	20g
土豆	Potato	50g
玉米	Corn	20g
土豆粉	Potato powder	50g
奶油	Cream	10g
盐	Salt	5g
芝士粉	Cheese powder	5g
胡椒	Pepper	4g
小番茄	Cherry tomato	10g
欧芹	Parsley	10g

温馨提示 (Kindly Reminder)

蟹肉也可以用冰鲜蟹肉代替。
(Crab meat can also be replaced with frozen crab meat.)

制作过程 (Method)

① 将所有原料准备齐全。
(Prepare all the ingredients.)

② 将蟹焯水，去壳，取肉。
(Blanch the crab, remove the shell and take out the meat.)

③ 用玉米、小番茄、欧芹、盐和胡椒调制玉米莎莎备用。
(Prepare salsa with corn, cherry tomatoes, parsley, salt and pepper.)

④ 将土豆煮烂后碾成泥，然后加入奶油、黄油、土豆粉、芝士粉、干葱和蟹肉，搅拌在一起搓成饼。
(Cooking potatoes in boiling water is a first step for making mashed potatoes. Then add the potato powder, cheese powder, crab meat, shallot and mix it together to form a meat cake.)

⑤ 将土豆饼用低油温慢慢煎熟。
(Pan-fry the meat cake with low heat.)

⑥ 最后配上玉米莎莎装盘即可。
(Garnish with salsa dressing at the end, then it is ready to serve.)

❶

❷

❸

❹

❺

❻

第9章 东南亚菜

越南春卷
Vietnamese Rice Paper Spring Rolls

菜肴特点
（Dish Characteristic）

口感清爽，简单味美
Fresh, simple and delicious

原料 (Ingredients)

春卷皮	Spring roll rice paper	10g
罗马生菜	Roman Lettuce	10g
蟹肉棒	Crab stick	10g
甜辣酱	Sweet chili sauce	20g
芒果	Mango	10g
粉丝	Mung bean vermicelli	30g
黄瓜	Cucumber	5g
胡萝卜	Carrot	100g
香菜	Coriander	5g

温馨提示 (Kindly Reminder)

蟹肉棒可以用虾代替。
(Crab sticks can be replaced by shrimp.)

制作过程 (Method)

① 将所有原料准备齐全并洗净。
(Prepare all the ingredients and wash them thoroughly.)

② 将黄瓜、胡萝卜切丝，芒果切条。
(Slice the cucumber and carrot, cut mango into strips.)

③ 将粉丝焯熟后过冷水备用。
(Blanch the mung bean vermicelli with ice water and set aside.)

④ 将春卷皮在水中浸泡 10 秒左右捞起。
(Soak rice papers into the water for about 10 seconds and take out.)

⑤ 将罗马生菜、香菜、蟹肉等包在春卷皮中。
(Wrap Roman Lettuce, coriander and crab meat with rice papers.)

⑥ 挤上甜辣酱，装盘即可。
(Add sweet chili sauce and it is ready to serve.)

❶

❷

❸

❹

❺

❻

泰式牛肉沙拉
Thai Beef Salad

菜肴特点
(Dish Characteristic)

营养健康，口感清爽
Fresh and healthy

原料 (Ingredients)

牛排	Beef rump steak	200g
柠檬	Lemon	50g
鱼露	Fish sauce	10g
香菜	Coriander	10g
尖椒	Red chilies	5g
香茅	Lemongrass	5g
小番茄	Cherry tomato	15g
黄瓜	Cucumber	15g
洋葱	Onion	10g
薄荷叶	Mint leaves	10g

温馨提示 (Kindly Reminder)

牛肉最好用新鲜的。
(Fresh beef is necessary.)

制作过程 (Method)

① 将所有原料准备齐全。
(Prepare all the ingredients.)

② 将牛排煎至五成熟。
(Pan grill beef rump steak until medium well.)

③ 煎好的牛排冷却后切成型。
(Set the steak aside for cooling down, then shape it with a knife.)

④ 将柠檬、鱼露、香菜末、尖椒末、香茅末调成泰汁。
(Whisk together lime juice, fish sauce, coriander, chili and lemongrass in a bowl.)

⑤ 将黄瓜、小番茄、洋葱切好后拌入牛肉和泰汁搅拌均匀。
(Mixed shredded cucumber, cherry tomato and onion together in another bowl, put the steak in and drizzle with the freshly made Thai dressing.)

⑥ 加些薄荷叶装盘即可。
(Add mint leaves and it is ready to serve.)

❶

❷

❸

❹

❺

❻

冬阴功汤
Tom Yum Kung

菜肴特点
（Dish Characteristic）

口感酸辣，鲜美可口
Spicy and sour flavors, delicious

原 料 (Ingredients)

海鲜	Seafood	10g
香茅	Lemongrass	10g
鱼露	Fish sauce	10g
柠檬	Lemon	20g
椰浆	Coconut cream	10g
青柠	Lime	30g
冬阴功酱	Tom Yum Paste	5g
小番茄	Cherry tomato	10g
桑巴酱	Sambal Sauce	10g
南姜	Galangal	10g
香菜	Coriander	5g
草菇	Straw mushrooms	10g
小米椒	Chili	5g

温馨提示 (Kindly Reminder)

从字面上看，"冬阴"来自两个泰语词汇。"冬"是煮的意思，"阴"是辣的意思，"功"是虾的意思，"冬阴功"是指一种辣的虾汤。
（Literally, the word "Tom yum" are derived from two Thai words: "tom" and "yum". "Tom" refers to boiling process (soup, in this case). "Yum" refers to a kind of Laos and Thai spicy and sour salad. Thus, "Tom Yum" is a Laos and Thai hot and sour soup. "Kung" refers to shrimp, which translates to spicy and sour shrimp soup.）

制作过程 (Method)

① 将所有原料准备齐全并洗净。
（Prepare all the ingredients.）

② 将南姜、香茅、香菜茎、小米椒切末，草菇开花刀，小番茄和青柠对切开备用。
（Chop the Galangal, lemongrass, coriander stem, chili, bloom-knifed the straw mushroom, halved the cherry tomatoes and lemon.）

③ 将海鲜焯水。
(Blanch the seafood.)

④ 在锅中入油炒小米椒末、香茅末、姜末、草菇，加入冬阴功酱和桑巴酱炒匀，然后加入椰浆、鱼露和海鲜，用大火煮沸后调小火，加入番茄和柠檬汁煮30秒起锅。
(In the pan, add oil and stir-fry chopped chili, lemongrass, ginger, mushroom, add the Tom Yum Paste and Sambal Sauce, then add the coconut cream, fish sauce, fish stock soup and seafood. Bring water to a boil, add tomato and lemon juice, Simmer for another 30 seconds.)

⑤ 将冬阴功汤起锅倒入碗中。
(Place the soup into the bowl.)

⑥ 最后加入香菜、青柠装盘即可。
(Finally garnish with coriander and lime to serve.)

❶

❷

❸

❹

❺

❻

拉萨汤
Laksa

菜肴特点
(Dish Characteristic)

辣，椰香味，咖喱味
Hot, coconut, curry flavor

　　拉萨汤被美国有线电视新闻网评为全球美食第七位。与重庆酸辣粉不同的是，拉萨汤是靠自然食材经过烹调带来的酸辣，让人食欲大开，根本停不下来！
　　CNN rated Laksa as the seventh most delicious food in the world. Unlike Chongqing spicy and sour noodles, Laksa is made of natural ingredients.It is so yummy that people can't stop eating.

拉萨汤

原料 (Ingredients)

虾	Prawn	10g
黄瓜	Cucumber	10g
咖喱酱	Curry sauce	10g
鱼丸	Fish balls	20g
桑巴酱	Sambal Sauce	10g
粉丝	Mung bean vermicelli	30g
椰浆	Coconut cream	5g
豆芽	Bean sprouts	10g
鹌鹑蛋	Quail egg	10g
鱼露	Fish sauce	5g
香菜	Coriander	5g
拉萨叶	Laksa Leaves	5g
油豆腐	Fried bean curd	10g

 温馨提示 (Kindly Reminder)

鹌鹑蛋可以用鸡蛋代替。
(Quail egg can be replaced by regular egg.)

制作过程 (Method)

① 将所有原料准备齐全并洗净。
(Prepare and clean all the ingredients.)

② 将黄瓜切丝，挑选香菜叶子备用。
(Shred the cucumber, preserve coriander leaves.)

③ 将锅中加入少许油、桑巴酱、咖喱酱、拉萨叶炒香，然后加入高汤煮开，加入虾、鹌鹑蛋、油豆腐、鱼丸煮20分钟，最后放入椰浆、鱼露调味。
(Place a large saucepan over the heat. Add oil, Sambal Sauce, curry sauce, Laksa Leaf and stir-fry until fragrant. Add the stock and bring to a boil then reduce heat to a simmer. Add prawn, quail eggs, fried bean curd and fish balls for 20 minutes.)

④ 起锅装碗，放入焯过水的粉丝、豆芽。
(In a bowl, place the blanched mung bean vermicelli and bean sprouts.)

⑤ 倒入拉萨汤、虾、鱼丸。
(Add Laksa soup, prawn, seafood fish balls.)

⑥ 最后放上黄瓜丝和香菜叶装盘即可。
(Finishing touch by placing cucumber slice and coriander leaves.)

❶

❷

❸

❹

❺

❻

印尼炒饭
Nasi Goreng

菜肴特点
(Dish Characteristic)

香辣可口，口味浓郁
Rice in spicy and savory

制作印尼炒饭很关键的一点就是桑巴酱。桑巴酱是一种以虾米、辣椒、糖等为原料烹制而成的酱料，有提鲜、增加风味的作用。

The key of making nasi goreng is all about its sauce. Samba Sauce is made from shrimp, chill peppers and sugar. It can increase freshness and flavor.

原料 (Ingredients)

米	Rice	100g
鸡蛋	Egg	50g
虾	Prawn	20g
洋葱	Onion	10g
虾片	Prawn cracker	20g
小米椒	Chili	5g
青豆	Green bean	10g
酱油	Soy sauce	5g
盐	Salt	5g
桑巴酱	Samba Sauce	10g
胡椒	Pepper	4g

温馨提示 (Kindly Reminder)

印尼炒饭可搭配沙爹鸡肉串、桑巴虾串、黄瓜、番茄等食物食用。

(Nasi goreng can be served with satay chicken, samba prawn, cucumber, tomato etc.)

制作过程 (Method)

① 将所有原料准备齐全。
(Prepare all the ingredients.)

② 将鸡蛋单面煎。
(Fry the eggs on one side.)

③ 将虾片炸脆当配菜备用。
(Fry prawn crackers as a side dish.)

④ 将米饭蒸熟后加入鸡蛋翻炒均匀起锅。
(Steam the rice. Heat the pan, and stir-fry the scramble eggs and rice together.)

⑤ 在热油锅中加入小米椒粒、洋葱碎、桑巴酱、酱油、盐、胡椒炒香,然后加入虾肉、青豆翻炒,最后加入炒过的蛋炒饭。
(Heat up the pan, stir-fry the chili, chopped onions, samba sauce, soy sauce, salt and pepper, then add in the shrimp, green beans and fried rice.)

⑥ 最后装盘时配上煎好的鸡蛋即可。
(Place the fried egg on top of the fried rice.)

❶

❷

❸

❹

❺

❻

马来姜蓉虾
Malaysian Prawn with Ginger

菜肴特点
(Dish Characteristic)

姜味浓郁，香辣可口
Rich in ginger, spicy and tasty

原料 (Ingredients)

虾	Prawn	100g
香菜	Coriander	10g
姜	Ginger	20g
酱油	Soy sauce	10g
大蒜	Garlic	20g
料酒	Rice wine	5g
香茅	Lemongrass	10g
葱	Shallot	5g
小米椒	Chili	5g

温馨提示 (Kindly Reminder)

如果喜欢吃辣，也可以把辣椒切成茸，与姜茸一起炒香。
(If you like spicy food, you can also add chopped chili with the ginger to pan-fry it.)

制作过程 (Method)

① 将所有原料准备齐全。
(Prepare all the ingredients.)

② 将虾对切开，去肠线。
(Halve the prawn and remove the black intestinal line.)

③ 将姜、香茅大蒜切成茸。
(Chop ginger, lemongrass and garlic.)

④ 将虾放入高油温中炸至金黄色。
(Fry the prawn in hot oil until golden brown.)

⑤ 将姜茸用油煸炒干后加入炸过的虾，烹入料酒和酱油调味。
(Stir-fry the ginger in oil and add the fried prawn. Cook with rice wine and soy sauce to taste.)

⑥ 最后加入葱段、小米椒、香茅和香菜装盘即可。
(Finally serve on a plate with spring shallot, chili, lemongrass and coriander.)

①

②

③

④

⑤

⑥

桑巴烤鱼
Grilled Sambal Fish

菜肴特点
(Dish Characteristic)

微辣，风味独特
Slightly spicy and unique flavor

说到桑巴，大家可能会想到热情洋溢的桑巴舞，那这道用桑巴酱制作的烤鱼相信也能带来热情似火的感觉。值得一提的是，这道菜用到的海鲈鱼的脂肪构成比淡水鲈鱼的更优质，油脂中的多不饱和脂肪酸含量更高。

When it comes to Samba, one might think of a passionate Samba Dance. This baked fish made with Sambar Sauce is believed to bring a feeling of passion as well. It's worth mentioning that this dish uses sea bass with a better fat composition and a higher polyunsaturated fatty acid content than freshwater sea bass.

桑巴烤鱼

原料 (Ingredients)

海鲈鱼	Sea bass	250g
柠檬	Lemon	10g
青柠	Lime	10g
小米椒	Chili	5g
桑巴酱	Sambal Sauce	100g
香菜	Coriander	5g
芭蕉叶	Banana leaves	5g

温馨提示 (Kindly Reminder)

鲈鱼最好选择新鲜的。
(Fresh sea bass is mostly preferred.)

制作过程 (Method)

① 将所有原料准备齐全并洗净。
(Prepare all the ingredients and wash them thoroughly.)

② 将海鲈鱼改刀成型。
(Shape and flatten the sea bass.)

③ 将鲈鱼肚子和表皮都抹上桑巴酱，然后放在芭蕉叶上面。
(Spread the Sambal Sauce on the belly and skin of the sea bass, then place it on top of the banana leaf.)

④ 将腌制好的海鲈鱼和芭蕉叶一起放入烤箱烤20分钟。
(Place marinated sea bass on banana leaf, put them into the oven for 20 minutes.)

⑤ 将小米椒和柠檬加工成型。
(Process chili and lemon into shape.)

⑥ 最后将小米椒、柠檬、青柠、香菜放在烤好的鱼上，装盘即可。
(Put chili, lemon, lime and coriander on the sea bass, and it is ready to serve.)

①

②

③

④

⑤

⑥

图书在版编目（CIP）数据

美味西餐 / 新东方烹饪教育组编 .—北京：中国人民大学出版社，2019.2
ISBN 978-7-300-26340-3

Ⅰ.①美… Ⅱ.①新… Ⅲ.①西式菜肴-菜谱 Ⅳ.① TS972.188

中国版本图书馆 CIP 数据核字（2018）第 236528 号

美味西餐

新东方烹饪教育　组编

Meiwei Xican

出版发行	中国人民大学出版社			
社　　址	北京中关村大街 31 号	邮政编码	100080	
电　　话	010-62511242（总编室）	010-62511770（质管部）		
	010-82501766（邮购部）	010-62514148（门市部）		
	010-62515195（发行公司）	010-62515275（盗版举报）		
网　　址	http://www.crup.com.cn			
经　　销	新华书店			
印　　刷	北京瑞禾彩色印刷有限公司			
规　　格	185mm×260mm　16 开本	版　次	2019 年 2 月第 1 版	
印　　张	9	印　次	2022 年 7 月第 3 次印刷	
字　　数	155000	定　价	36.00 元	

版权所有　侵权必究　印装差错　负责调换